AF521539

THE CONCEPT OF ŚŪNYA

THE CONCEPT OF ŚŪNYA

Editors
A.K. BAG
and
S.R. SARMA

INDIRA GANDHI NATIONAL CENTRE FOR THE ARTS
INDIAN NATIONAL SCIENCE ACADEMY
ARYAN BOOKS INTERNATIONAL

THE CONCEPT OF ŚŪNYA

ISBN: 978-81-7305-240-8

© **Indira Gandhi National Centre for the Arts, New Delhi**
and
Indian National Science Academy, New Delhi

No reproduction or translation of this book or part thereof in any form, except brief quotations, should be made without the written permission of the copyright holder and the publishers. The views expressed in this volume are those of the authors and are not necessarily those of the editors or the publishers

First published in **2003** by

INDIRA GANDHI NATIONAL CENTRE FOR THE ARTS
Central Vista Mess, Janpath, New Delhi-110 001 (India)

INDIAN NATIONAL SCIENCE ACADEMY
Bahadur Shah Zafar Marg, New Delhi-110 002 (India)

and

ARYAN BOOKS INTERNATIONAL
Pooja Apts., 4B, Ansari Road, Darya Ganj, New Delhi-110 002 (India)
Tel. : 23255799, 41099562; E-mail : aryanbooks@gmail.com

Computer Typeset at ABI Prints & Publishing Co., New Delhi
Printed in India by Replika Press Pvt. Ltd.

Contents

Ramification of Śūnya in Other Fields

Introduction

The concept of *śūnya* in India has a long history and varied manifestations in different dimensions in speculative thought in mathematics, philosophy and in mysticism. The Sanskrit term *śūnya* is used both as substansive and adjective, with diverse connotations, such as absence (*lopa*), universe (*ākāśa*), emptiness (*śūnyatā*), dot (*bindu*), circle (*śūnya)* etc. In mathematical literature it is used in the sense of 'zero' to fill up the blank space having no substantial numerical value of its own but playing a key role in the system of decimal notation, to express all numbers with nine digits, one (1) to nine (9) and the *śūnya* as the tenth. The application of *śūnya* in this system of notation was discovered in India, sometimes in the pre-Christian era. Its concretisation in the form of a circle or dot and its use in decimal place value became evident in the fifth century A.D.

No one contests the fact that the decimal system came out of India and it travelled slowly westwards, penetrating the lands where sexagesimal system was used, western Asia, North Africa and Europe. The whole world gradually recognised it as the most scientific system of numeration. Though erroneously called by the westerners as the Arabic numeral systems, now the entire world knows that the credit for discovery of *śūnya* goes to India. Medieval Central Asian scholars such as Al-Khwārizmī (9th century A.D.) and Al-Bīrūnī (11th century A.D.), George Sarton and other experts of modern time bear testimony that the system consisting of nine numerical digits and one zero *śūnya* was invented by the Hindus, i.e. Indians.

The Babylonian and Chinese records of about 200 B.C. and 3rd century A.D. show the use of blank space to denote the absence of a number, but it did not form the basis of their systems, as the Indian *śūnya* does. The question of India's indebtness to other cultures in this respect can safely be ruled out.

The treatment of all the fundamental operations with *śūnya* in mathematics and astronomy in a logical and systematic way was developed by subsequent authorities from Brahmagupta of the seventh century to Raṅganātha of the seventeenth century. In the way *śūnya* (zero) is defined as an infinitesimal, and this is exactly the same as the modern conception of zero in arithmetic.

Besides in mathematics and astronomy, the term, *śūnya* has been widely used to denote divergent concepts of Indian philosophical and mystical traditions. It denotes Brahman on one side and void, or absence of all phenomena, both material and abstract, on the other. It is the kernal of *Śūnyavāda* philosophy of the Mādhyamika Buddhists where *śūnya* is equivalent to emancipation (*Nirvāṇa*), transcending all four dimensions of existence and non-existence along with their permutations and computations. In Tantra-Āgama traditions, particularly in the Śaivāgamas, the concept of *śūnya* and *śūnyatā* have several levels, starting from void or emptiness in place and time, to ultimate and absolute bliss. In the Vedāntic view *śūnya* or *avyakta* or *nirguṇa* are collinear terms which negate empirical reality of the phenomenal world.

It is also believed that the philosophical conceptualization of *śūnya* and *śūnyata* as developed by Nāgārajuna and of the *ākāśa* of Kaṇāda lent support to and made an impact on the concretization of *śūnya* of the mathematicians. At the same time the concept of *śūnya* of the philosophers and Tantrics has been symbolized in the temple architecture and iconography.

The present seminar was jointly organised by the Indian National Science Academy (INSA) and the Indira Gandhi National Centre for the Arts (IGNCA) under the convenorship of Dr. Kapila Vatsyayan and its major aim was to collect a documented account of various facets on the notion of *śūnya* on the three areas, viz. *śūnya* in speculative thought, *śūnya* in mathematical systems, and ramification of *śūnya* in other fields (Temple Art, Tantra and Agama, Mysticism etc.). The order of presentation of papers has been slightly changed in order to give to it a thematic linkage. A few papers could not be accomodated.

About sixty experts attended the seminar. Three foreigners, one from China and two from USA, have also contributed. It was indeed a lively forum for a free and candid exchange of ideas, besides many thoughtful papers presented and discussed. The participants were of the general opinion that the seminar was academically fruitful, and possibly helped for better appreciations of the problem.

We are grateful to the IGNCA for entrusting to us the responsibility of editing the proceedings of the seminar. It is a pleasure to thank Dr. Kapila Vatsyayan, former Academic Director, IGNCA, and Dr. S. Varadarajan, former President, INSA, for their scholarly guidance. We are highly obliged to Pandit Satkari Mukhopadhyaya, former Coordinator, *Kalā-kośa*, IGNCA, for his valuable help in the preparation of this volume. We also wish to thank Swami Jitatmananda, Prof. J.N. Kapur, Prof. Vidya Niwas Mishra and Prof. Lokesh Chandra who presided over the various sessions of the seminar, and to all the scholars who participated in the seminar for their valuable contributions. Finally, we wish to record our appreciation of Dr. Sudhir Kumar Lall of the IGNCA for his able assistance in seeing the volume through the press.

It is hoped that the proceedings, though delayed, will be a useful addition to the existing literature on the subject.

Editors

The Contributors

ADHIKARI, M.R.
Department of Mathematics, University of Burdwan, Burdwan - 713104.

BAG, A.K.
Adviser, History of Science Programme, Indian National Science Academy, Bahadur Shah Zafar Marg, New Delhi - 110002.

CHAKRAVARTY, H.N.
Indira Gandhi National Centre for the Arts, B1/3, E-7, Lane No. 2, Ravindra Puri Extension, Varanasi - 221005.

DESAI, DEVANGANA
Editor, *Journal of the Asiatic Society of Bombay*; Shanti 1/30, 19 Pedder Road, Mumbai - 400026.

GUPTA, R.C.
Former Professor of Mathematics, BIT (Meshra); Ras Bahar Colony, P.O. Lahar Gird, Jhansi - 284003.

JITATMANANDA, SWAMI
Shri Ramakrishna Ashram, Dr. Yagnik Road, Rajkot - 360001.

JOSHI, M.C.
Former Member - Secretary, Indira Gandhi National Centre for the Arts, Janpath, New Delhi - 110001.

KAPUR, J.N.
Former VC, Meerut University; C-766, New Friends Colony, New Delhi - 110068.

MALIK, S.C.
Professorial Research Scientist, Nehru Memorial Museum and Library; 185, Mandakini Enclave, New Delhi - 110019.

MANANSALA, P.
3633. Comanche way, Antelope, CA 96843, USA.

MUKHERJEE, B.N.
Former Carmichael Professor of Ancient History and Culture, Calcutta University; 56, Jatin Das Road, Calcutta - 700029.

MUKHOPADHYAYA, A.
Flat C-2, Kallol, No. 10, T.N. Chatterji Street, Calcutta - 700090.

PANDIT, M.D.
Centre for Advanced Studies in Sanskrit, University of Poona, Pune - 411007.

PARAMHANS, S.A.
Department of Applied Mathematics, Institute of Technology, Banaras Hindu University, Varanasi - 221005.

PINGREE, DAVID
Professor of History of Mathematics, Box 1900, Brown University, Providence, Rhode Island 02912, USA.

RAJU, C.K.
C-17, Oxford Apartment, I.P. Extension, Delhi - 110092.

SARMA, K.V.
Former Director, VVRI Institute (Hoshiarpur); East Coast Flats, 32/2, II Main Road, Gandhi Nagar, Adyar, Chennai - 600020.

SARMA, S.A.S.
Department of Sanskrit, French Institute of Indology, 11, St. Lous St., P.B. 33, Pondicherry- 605001.

SARMA, SREERAMULA RAJESWARA
Former Professor of Sanskrit, AMU; 23, Safina Apartments, Medical College Road, Aligarh - 202002.

SATHYANARAYANA, R.
Trayee Lakshmi, 9th Cross, Jayanagar, Mysore - 570014.

SHARMA, PREM LATA

Former Vice-chancellor, Indira Sangita Kala Vishvavidyalaya, Khairagarh; Bharat Nidhi, Amnaya, 209/1, Near Nand Nagar, Varanasi - 221005.

SHASTRI, AJAY MITRA

Former Professor of Ancient Indian History, Culture and Archaeology, Nagpur University; Prachi, 23, Vidya Vihar, Rana Pratap Nagar, Nagpur - 440022.

YUSHENG, WANG

Institute for History of Natural Science, CAS, 137 Chao Nei Da Jie, Beijing 100010, China.

Śūnya in Speculative Thought

Śūnyatā : Its Meaning in Philosophy and Science

SWAMI JITATMANANDA

The Upaniṣads, the crest jewel of Vedic wisdom, begin and end with the well-known invocation:

om pūrṇam adaḥ pūrṇam idam pūrṇāt pūrṇam udacyate /
pūrṇasya pūrṇam ādāya pūrṇam eva avaśiṣyate //

"Om, the infinite fullness is that Brahman. The same infinite fullness is this Brahman. The same infinite fullness springs from (is hidden) in this finite universe. When this same infinite fullness is taken away from the infinite fullness, still what remains is infinite fullness".

The Vedic seers realised the Ultimate Reality or Brahman as Infinite knowledge, Existence and Bliss. Brahman is real (*brahma satyam*), Śaṅkarāchārya, the greatest champion of Advaita Vedānta asserted. Only, he continued, in order to attain that realisation one must realise this transient phenomenal universe as a void, a total emptiness — *jagat mithyā.*

After realisation, this empty, transient evanescent world of space–time appears, as a manifestation of that One Reality Brahman (*jīvo brahma eva na aparaḥ*). Nature abhors vacuum. Voidness is always filled with fullness.

The realisation of *śūnyatā,* void, or emptiness of this space–time existence is the essential step to the final bliss of Infinite fullness, the Brahman beyond space–time. The Ṛg-vedic poem on Non-Existence, *Nāsadīya Sūkta,* speaks about this state of void:

"Existence was not there, nor non-existence. The world was not, the sky beyond was neither. What covered the mist? Of whom was that, what was in the depths of darkness thick? He who presides over the highest heaven, He may perchance, know it, or even He may not know".

Voidness or *śūnyatā* is a common experience of all seekers of God at the penultimate stage of realisation. Shri Ramakrishna had this realisation: "Divine mother revealed to me that everything is void".[1]

"Put 50 zeros after 1, large sum. Erase 1 and nothing remains. First one then many" says Shri Ramakrishna.[2]

Shri Ramakrishna could take frequent flights to the world of super consciousness beyond space–time. Buddha's experience was beyond both existence and non-existence. "He is not an atheist, he simply could not express inner experiences in words", said Shri Ramakrishna.[3]

"Meditation on and becoming one with pure intelligence is Buddha", says Shri Ramakrishna.[4]

"The Letter Killeth", said Saint Paul about expressing the ineffable, ultimate Reality.

The metaphysical concept of *śūnyatā* or void is paradoxically an offshoot of the Mahayanic Buddhism which virtually brought "a radical revolution" —transforming Buddhism with "a new spirit", freeing it from the narrow monasticism oriented religion contemplating on annihilation and transience of life, to a dynamic and expanding world religion with its highly positive ideas of salvation by the grace of Bodhisattvas (Buddhas–to–be) "a salvation not in annihilation but in eternal life" its *pāramita* (perfections in life),[5] and *dharmakāya* (the body of Being) and above all a supremely living personality of Buddha as the goal of all spiritual achievement (*Buddhaṃ śaraṇaṃ gacchāmi*). "*Nirvāṇa* is in itself a supremely positive experience", and "the *dharmakāya* is both Supra-personal Absolute beyond the range of thought and the supreme, all-embracing personality revealed not only in the historical wisdom of the Buddhas, but in their boundless love and compassion", writes Sidney Spencers.[6]

Śūnyatā is the other side of Pūrṇatā : The Upaniṣads (Emptiness is the other face of Fullness)

The Ultimate Reality in the Upaniṣads is *"not this, not this"*. The *Bṛhadāraṇyaka Upaniṣad* describes the Ultimate Reality or Brahman as *neti neti*. "Now is given the final definition of Brahman — *Not this, Not this;* No greater definition is there than this one" (*"athātaḥ ādeśaḥ neti neti, na hi etasmātiti. na iti anyat param asti — Bṛ. Up.* 2.3.6)

or

Brahman is that from which words or ordinary minds return :

yato vāco nivartante aprāpya manasā saha.

(Taittirīya Up. 2.9)

The Upaniṣad, again, describes this indescribable Brahman as "Full with all the powers, knowledge and bliss".

(Ātman), the individual self, is one with Brahman which is full of all knowledge, all meditation, all energy, all space, all light, all the universe.

sa va ayam ātmā brahma, vijñānamayo, manomayaḥ, prāṇamayo, ākāśamayaḥ, tejomayaḥ, pṛthivimayaḥ.

(Bṛ. Up. 4.4.5)

or

In the beginning there was this Brahman alone. It knew itself as *I am Brahman* :

brahma vā idam agra āsīt tad ātmānam evāvet ahaṃ brahmāsmīti.

(Bṛ. Up. 1.4.10)

or

"O Gārgi, by this imperishable Brahman the entire universe and space are interpenetrated.

sa khalu akṣare gārgi otaś ca protaś ca.

(Bṛ. Up. 3.8.11)

or

Turīya (The transcendent beyond Space-Time, The unspeakable Reality) is held to be the all-pervasive source of all entities.

(Gauḍapāda's comment on *Māṇḍūkya Upaniṣad; verse 10*)

or

Brahman is Truth, Knowledge and Infinity.

satyaṃ jñānam anantaṃ brahma.

(*Tait. Up.* 2.1.1)

or

The self is permeated with cosmic energy, universal mind, Absolute knowledge and Absolute bliss.

The Human body is filled with this Self (*tenaiṣa pūrṇam*). The self is:

prāṇamaya, manomaya, vijñānamaya, ānandamaya.

(*Tait. Up.* 4)

or

For whenever an aspirant gets fearlessly established in this unperceivable, bodiless, inexpressible and unsupporting Brahman, he reaches the state of fearlessness:

yadā hi eva eṣa etasmin adṛśye, anātmye, anirukte, anilayane, abhayaṃ pratiṣṭhāṃ vindate, atha saḥ abhayaṃ gato bhavati.

(*Tait. Up.* 2.7.1)

or

Know that *ātman* alone which has interpenetrated this heaven, earth, the internal and external universes including mind and all the vital energies. Give up all other vain talks. For that Ātman alone is the bridge from this relative existence to immortality.

yasmin dyauḥ prithivī cā āntarikṣam otam, manaḥ saha prāṇaiś ca sarvaiḥ /
tam evaikaṃ jānatha ātmānam anyā vāco vimuñcathā amṛtasyaiṣa setuḥ //

(*Muṇḍaka Upaniṣad* 2.5)

Śūnyatā and Buddhism

The philosophical concept of *śūnya* or Voidness is primarily a product of Buddhistic philosophy and metaphysics which is divided into two important sects, viz. Hīnayāna (lower vehicle) and Mahāyāna (greater vehicle). Vaibhāṣika (Sarvāstivāda), Sautrāntika, Mādhyamika (Śūnyavāda) and Vijñānavāda are the four famous schools. The former two belong to Hīnayāna and the latter two to Mahāyāna. While the former two accept the reality of the external world, the latter two deny it.

The Mādhyamikas were so called on account of the emphasis they laid on *madhyama-pratipat* (the middle view). In his first sermon at Benaras, the Buddha preached the middle path which is neither self-mortification nor a life devoted to the pleasures of the senses. The middle path, as advocated by the later adherents of the Mādhyamika system, is not the same. It is more metaphysical than experiential or ethical as preached by Buddha.[7]

Here, the middle path stands for the non-acceptance of the two views concerning existence and non-existence, eternity and non-eternity, self and non-self and so on. *In short, it advocates neither the theory of reality nor that of the unreality of the world, but merely of relativity.*

Śūnyavāda originated from the times of Buddha himself who always was silent about God or Soul. Thus, we find Subhūti saying to the Buddha that all *skandhas* (or groupings of the internal world) like *vedanā* (feeling), *saṃjñā* (concepts) and the *saṃskāras* (conformations) are all *māyā* (illusion). All *skandhas* (aggregates), *dhātus* (elements) and *āyatanas* (forms) are void and absolute cessation. The knowledge of everything phenomenal as pure void is regarded as the highest knowledge (*prajñāpāramitā*).

The truth is neither eternal (*śāśvata*) nor non-eternal (*aśāśvata*) but pure void. It should be the object of a saint's endeavour to put himself in the "Suchness" (*tathatā*) and consider all things as *void*. Achievement of "Suchness" is *Prajñāpāramitā* (perfection in knowledge) which leads to *śīlapāramitā* (perfection in behaviour), and other *pāramitās* or *perfections* of Buddhist saints. The true *prajñāpāramitā* is the absolute cessation of all appearance (*yaḥ anupalambhaḥ sarvadhayamanam sā prajñāpāramitā ityucyate*).[8]

All phenomena, both being and non-being are illusory (*sadāsaniah māyopamāḥ). The* ultimate truth is that in which the positive and negative are one and the same (*bhāvābhāvasamānatā*). Such a state which is complete in itself and has no name and no substance had been described in the *Laṅkāvatāra-sūtra* as *suchness or thatness* (*tathatā*).[9]

This state of "suchness" is also described in another place in the *Laṅkāvatāra-sūtra* as voidness (*śūnyatā*) which has no origination and no essence. In another place, it is also designated as (*tathāgatagarbha*). In the *History of Indian Philosophy*, Shri Surendra Nath Dasgupta explains the various concepts of *śūnyatā* in Buddhism.[10]

The nothingness (*śūnyatā*) of things may be viewed from seven aspects. In the ultimate analysis Mādhyamika concept of *Śūnyatā* stands thus : (i) all things have their own nature being undeterminded; a reference not definable (*lakṣmanāśūnyatā*); (ii) that they have no positive essence (*bhāvasvabhāvaśūnyatā*) since they spring up from a natural non-existence (*svabhāvāvabhavoāpatti*); (iii) that they are of an unknown type of non-existence (*apracāritaśūnyatā*), since all the *skandhas* (groupings of feelings etc.) vanish in the *nirvāṇa*; (iv) that they appear phenomenally as causally connected though they are really non-existent (*pracaritaśūnyatā*); (v) that none of the things can be described as having any definite nature, which can be demonstrated by language (*nirabhilapsyaśūnyatā*); (vi) that there cannot be any knowledge about them except that which is brought about by the long-standing defects of desires (*vāsanā*) which pollute our vision; and (vii) that things are also non-existent in the sense that we affirm them to be in a particular place and time in which they are not (*itaretaraśūnyatā*).[11]

Aśvaghoṣa of the first century A.D. spelt out the fundamental thoughts of Mahāyāna Buddhism — in particular those relating to the Buddhist concept of "suchness" in a small book called the *Awakening of Faith.*[12] His works influenced the eminent Mahāyāna philosopher Nāgārjuna who with his brilliant arguments demolished the metaphysical ideas of his time and thus showed that reality, ultimately, cannot be grasped with concepts and ideas. Hence, he gave it the name *śūnyatā,* the "void", or "emptiness", a term which is equivalent to Aśavaghoṣa's *tathatā*, or "suchness", when the futility of all conceptual thinking is recognised and, reality is experienced as pure *suchness.*[13]

Sautrāntikas, the atomic realists among Buddhists, reduced the 75 elements of the Vaibhāṣikas into 43 which they believe exist objectively and independently, and rejected the rest as subjective and unreal. Nāgārjuna rejected these ideas and asserted that there is nothing which exists absolutely and objectively. Everything that exists is relative and dependent and, therefore unreal. Everything in this phenomenal world is relative to everything else.

The ground phenomena of this evanescent relative phenomenal world can never be understood by mere reason as reason by its very nature leads to insoluble antinomies. Entire phenomenal world is subjective; and the subject itself is unreal. It lacks any basis. Nāgārjuna thus comes to the conclusion that the Absolute Reality is *śūnya*. It is *śūnya* in the sense that it is transcendent to thought and words. It is indescribable, non-determinate and non-dual. It is neither subject, nor object.

The emptiness of Nāgārjuna, however, is far from being a nihilist statement. It only asserts like the Upaniṣads that all concepts about reality formed by the limited human mind are ultimately void. *śūnyatā* or emptiness is not nothingness. It is also the source of all life and essence of all forms.

That is why the essential nature of all things is described in Mahāyāna Buddhism not only by the abstract metaphysical as *Suchness* and *Void*, but also by the term *dharmakāya*, the "*Body of Dharma or Universal laws*" both physical and metaphysical. The *dharmakāya* is in a way similar to the *Brahman* of the Upaniṣads. It pervades all material things in the universe and is reflected in the human mind as *bodhi (knowledge* or enlightened wisdom). It is the spiritual entity that interpenetrates the material universe.[14]

Because of the positive connotation of "suchness" or "*śūnyatā*", emphasis on love and compassion as essential parts of wisdom has found its strongest expression in the ideal of the Bodhisattva. A Bodhisattva is a highly evolved human being on the way to becoming a Buddha, who is not merely contemplating on voidness or seeking enlightenment for himself alone, but has vowed to help all other beings achieve Buddhahood before he enters into *nirvāṇa*.[15]

The Chinese, like the Indians, believed that there is an ultimate reality which they called Tao which underlies and unifies the multiple things and events. Tao is "complete", all-embracing", "the whole". Tao, originally meant "the way". It is the way, or process of the universe, the order of nature.[16]

The *Mādhyamika-kārikā* deals with two kinds of truths: *saṃvṛti* (conventional or empirical truth) and *paramārtha* (higher or transcendent truth). The former refers to ignorance or delusion which envelops reality and gives a false impression, while the latter is the realisation that worldly things are non-existent like an illusion of an echo. It is seeker's endeavour to be established in this realisation which is perfect knowledge or *Prajñāpāramitā*.[17]

The Yogācāra school brings out the practical side of philosophy, while Vijñānavāda brings out its speculative features. The *Laṅkāvatāra-sūtra*, an important work of this school, maintains that only the mind (*cittamātra*) is real, while external objects are not.

The Yogācāra differs from the Mādhyamika only in that it attributes qualities to ultimate Reality. It also holds that reality is pure consciousness *(vijñānamātra)* while the Mādhyamikas believe it as *śūnyatā.*

All concepts like *tathatā, or bhūtatathatā* (suchness) are twofold, having a negative and positive aspect and just as *śūnyatā* is negatively the absence of particularity, it is positively the apotheosis of the Thervāda doctrine of *aniccata* (*anityatā*) change, the eternal flux of becoming. "It was Buddhist philosophy which formed the theory, but that which supplied it with facts to substantiate it was Buddhist experience", says Christmas Humphreys.[18]

Asaṅga in his treatise known as *Mahāyānasūtrālaṅkāra* adopts a more positive concept of Reality than Nāgārjuna, which is known as Vijñānavāda. He does not accept the claim of Nāgārjuna that the phenomenal world is merely conceptual without any basis. Unlike the *Śūnyavāda* of Mādhyamika school he thought that the changing, unreal phenomena must be rooted in some stable reality which, according to him, is pure consciousness.

Śūnyavādins uphold that the consciousness is an aggregate of momentary consciousness. Asaṅga does not treat the momentary consciousness as the sole reality. Reality is non-dual, free from subject-object duality, indescribable and beyond the grasp of the intellect. It can be realised intuitively. Momentariness applies only to changing phenomena. Asanga emphasises the absoluteness of pure consciousness (*vijñaptimātra*) by such words as *dhruva and nitya— unchanging, permanent* and *eternal*. His notion of Absolute Reality clearly show that Asaṅga's Vijñānavāda cannot be termed as subjective idealism. It is more akin to Vedantic monism.

Asaṅga calls the experience of this absolute reality by different names *Nirvāṇa, Dharmadhātu, Viśuddha Tathatā, Tattva, Parābodhi, Buddhatva* etc. *Nirvāṇa* is beyond all thought and discrimination. It is indescribable and indefinable. That is why Asaṅga calls *Nirvāṇa* as *Vikalpopaśama,* i.e. destruction of imaginary thoughts and things inspired by ego. Asaṅga asserts that the ever-changing ego of the individual is not the eternal Ātman.[19]

Asaṅga glorifies the Vijñānavāda because of the greatness of the support *(ālambana)* of vast expanse of Buddhist sūtra literature, (ii) of the greatness of understanding for oneself and making others understand it, (iii) of the greatness of two kinds of knowledge, viz. *Pudgalanairātmya* (non-substantiality of self) and *dharma-nairātmya* (non-substantiality of elements), (iv) of the greatness of doing good deeds, constantly at all times, (v) of the greatness of not taking a nihilistic approach to life on account of sufferings, (vi) of the greatness of complete knowledge of perfect Buddha and (vii) of the greatness of the activities of Buddhas and seeking, again and again, the final enlightenment.[20]

Śaṅkara, Mādhava and Rāmānuja refute *Śūnyavāda*. In his commentary on *Brahma Sūtra* (2:2:18) (*samudaya ubhayahetuke, itadaprātiḥ*) Śaṅkaracarya refutes the *Sauttrāntika* and the *Vaibhāṣika* schools of Hīnayāna Buddhism which believe (i) in the reality of every object or Sarvāstivāda; and the two schools of Mahāyāna Buddhism, namely (ii) Vijñānavāda which opines that conscious thought alone is real and the (iii) Śūnyavādins for whom everything is void or unreal.

The realists who do not believe in the existence of God or Brahman assume, says Śaṅkara, that the external world of elements, sense-organs and qualities and the internal world are both real. The external world arises out of four kinds of atoms, earth, water, fire and air. The inner world consists of five groups or *skandhas*: *rūpa* (sensation), *vijñāna* (knowledge), *vedanā* (feeling), *samjñā* (verbal knowledge) and *saṃskāra* (disposition). Neither the atoms nor the *skandhas* can by themselves achieve the groupings as assumed by the realists. They are non-intelligent. If they are assumed to be active of their own nature, they will be always active and there will be no scope for release. Some superior force or consciousness is required to aggregate them.[21]

Second, if according to Śūnyavādins everything is void and unborn how can something which has not yet obtained existence enter into connection with something else as they believe in their theory of *Pratītya-Samutpāda*?

Connection is possible between two existing things only, and not one existing and one non-existent thing or of two non-existing things. Only existing things can be spoken of as having limitations. Absolute non-existence or what is altogether featureless cannot be spoken of as "being prior to" origination.

Śaṅkara refers to the *Chāndogya Upaniṣad*:

brahma evedam agra āsīt; ekam eva advitīyam.

(Ch. Up. 6.1.1)

"In the beginning there was Being alone, one without a second". This experiential knowledge of *śruti*, according to Śaṅkara, refutes the suggestion of the non-existence as the source of temporary existence on the ground that the existent cannot come out of the non-existent.

In the *Brahma Sūtra* (2.2.31) *kṣaṇikatvāc ca*, *Śaṅkara* argues that Śūnyavāda or complete denial of everything is not possible. Without some permanent background or truth in whose perceptive travience of realistic can be felt. Swami Vivekananda nicely argues that fleeting shows of the phenomenal realistic can be understood only in the background of a permanent screen or background which is Brahman.

The other sūtra (2.2.32), "*sarvathā anupapatteḥ ca*" again refutes *Śūnyavāda* (and on account of being defective in all ways). Rāmānuja takes this sūtra to refute the *Śūnyavāda* that nothing whatever is real. The world we know is mutable. Its events lapse into non-entity at the very moment of their birth. The world is *saṃsāra*, a perpetual transience of things, devoid of any substance. The objects of empirical knowledge are unstable, transient, continuously transforming by new relations to other things, which, when scrutinised, prove equally relative and elusive. This imperfect world, according to Rāmānuja, makes us realise the Real Being which underlies this unstable world. Otherwise how would a Buddha come out of mere ignorance due to momentariness and transience of life?

Mādhava looks upon Śaṅkara's Advaitism (that the Ultimate Reality is *neti neti*) as crypto-Buddhism. There is no difference, he asserts, between the unqualified Brahman or all pervading consciousness that is beyond words and thoughts and the *śūnya* of the Mādhyamika system. The non-dualists treat it as unspeakable and unknowable, at the same time they refer all knowledge to it.[22]

Srinivasa says that the view of *Śūnyavāda* is faulty since it rejects Brahman as the fundamental substratum of life. It admits the collections of atoms, unseen and unheard, but says that a cause for their collection is an absurdity.[23]

Sāṃkhya holds that there can be no production of a thing previously non-existent. This doctrine is called Satkāryavāda, i.e. that the *kārya* or effect is *sat* or existent even before the causal operation to produce the effect was launched.[24]

Vivekananda: Experience of Void and Rejection of Śūnyavāda

As a young seeker of God, Vivekananda asked an eminent scholar, "The author of *Pañcadaśī* also says, 'What we call Brahman is the same truth as the *śūnya* of the Buddhist.' What does all this mean?"[25]

During later years, his own experience of the Ultimate made him sing in his own song of *Samādhi* :

"Void merged in void.
The unspeakble, the inexpressible !
He only understands who has realised THIS !"

The famous lines of the Upaniṣad always brought back the ecstasy of his own realisation of the unspeakable !

na tatra sūryo bhāti na candratārakam, nemā vidyuto bhānti kuto'yam agniḥ /
tam eva bhāntam, anubhāti sarvaṃ, tasya bhāsā sarvam idaṃ vibhāti //

— *Kaṭha Up.* 2.2.15.

"There the sun shines not, neither the moon nor the stars, nor lightning is there. What to speak of fire? Yet by its efflorescence all these universes radiate".

In the language of reason he explained *nirvāṇa* in the West : "The Jñānī takes nothing for granted. He analyses by pure reason and force of will, until he reaches *nirvāṇa* which is the extinction of all relativity. No description or even conception of this state is possible".[26] Or "The Jñānī seeks to tear himself away from this bondage of matter by the force of his intellectual conviction. This is the negative way of the *Neti Neti* - not this, not this".[27]

One remembers that Buddha's answers to question about God was silence. In his lectures on Jñāna Yoga, Vivekananda first expresses the Buddhist stand-point. "Nobody has ever seen a substance, none can ever conceive; what is the use of thinking about this substance? Why not become a Kṣanikavādin and say that whatever exists is this succession of mental currents and nothing more?

"What is your individuality? You have one thought tonight and another tomorrow... And the Buddhists argue that along with this idea of a Personal God comes tyranny and priestcraft.... The Buddhists may even urge that, that is why he is going to create annihilation, *nirvāṇa*, to escape this".[28]

Vivekananda, then, refutes the Buddhist objections and by the same Buddhistic logic establishes the existence of One all pervading reality. The first Buddhisitic objection, the metaphysical one, is that substance and qualities are different. The Advaitist says that they are not. "There is no difference between substance and qualities. You know the old illustration, how the rope is taken for the snake and when you see the snake you do not see rope at all, the rope has vanished".

"Dividing the thing into substance and quality is a metaphysical something in the brains of philosophers, for never can they be in effect outside".[29]

"If substance is unqualified, there can only be one. If you take qualities off from the soul, and show that these qualities are in the mind really, super-imposed on the soul, then there can never be two souls, for it is qualification that makes the difference between one soul and another. How do you know that one soul is different from the other? Owing to certain differentiating marks, certain qualities. And where qualities do not exist, how can there be differentiation? Therefore, there are not two souls, there is but One. That one is called Paramātman, that very One is called Jivātman and so on".

That is why the Vedantist says so long as there are two, fear does not cease. Fear comes from two (*dvitīyāt vai bhayam*; *Br. Up.*) It is only when one does not see another, does not feel another, when it is all one — then alone fear ceases, then alone death vanishes, then alone *saṃsāra* vanishes.

Vivekananda refutes the second Buddhist objection by the simple logic of relativity of motion. The second objection is that the universe is in continuous motion. It is the philosophy of *kṣaṇikavādins,* believers in momentariness of all phenomena. Vivekananda argues that in this little world of ours, the *jagat*, thought and feeling, mind and body, men and animals and plants are in a continuous state of flux. But suppose you take the universe as a unit whole; can it change or move? Certainly not. Motion is possible in comparison with something which is a little less in motion or entirely motionless. When the realisation of "I am the universe" comes, all fear ceases.[30]

Modern Physics: A Holistic Universe

Matter and empty space were the two fundamentally distinct concepts on which the atomism of Democritus and of Newton was based. In general relativity, if there is a massive body, there will also be a gravitational field, and this field will manifest itself as the curvature of the space surrounding that body. In the cosmic universe everything, emptiness and matter are inseparably related.

Relative Physics shows us that material objects are not distinct entities, but are inseparably linked to their environment or the so-called empty space; properties of one material substance can only be understood in terms of their interaction with the rest of the world. According to the principle of Ernst Mach this interaction reaches out to the entire universe, to the distant stars and galaxies. The basic unity of the cosmos manifests itself, therefore, in the macrocosmic world of the very large. Astronomer Fred Hoyle says the unity and interrelation between a material object and its environment, appears in an even more striking form at the sub-atomic level.[31]

The classical field theory, electrodynamics, has been merged with quantum theory into a theory called quantum electrodynamics which describes all electromagnetic interactions between sub-atomic particles. It is a world where the unit of time according to Max Plank is 10-43 sec. and the unit of space is 10-35 metres.

This theory incorporated both quantum theory and relativity theory. It was the first 'quantum-relativistic' model of modern physics and is still the most successful. Here the primary concept is of a 'quantum field', that is, of a field which can take the form of quanta, or particles.

This is indeed an entirely new concept which has been extended to describe all sub-atomic particles and their interactions. *In this new 'field theories', the classical contrast between the solid particles and the empty space surrounding them is eliminated.* The quantum field is seen as the fundamental physical entity. It is the continuous medium which is present everywhere. Particles are merely temporary condensation of the energy field. These are concentrations of energy which are born, annihilated and reborn into new type of particles at an incredibly short period of time, say a few particle seconds (a particle second is 10-23 seconds). The individual character dissolves into the underlying field.[32]

In the words of Albert Einstein, "We may therefore regard matter as being constituted by the regions of space in which the field is extremely intense .... There is no place in this new kind of physics both for the field and matter, for the field is the only reality".[33]

The Upaniṣads consider Brahman, Infinite beyond space and time, the underlying entity, as the only reality; all its phenomenal manifestations in space time are seen as transitory and illusory. Brahman resembles the quantum field of the physicist, because it is seen as the essence of all phenomena in this world and, consequently, is beyond space, time and all concepts and ideas. The quantum field, of modern physics, is a well-defined concept which only accounts for some of the physical phenomena at the sub-atomic level.

"Nevertheless, the intuition behind the physicist's interpretation of the sub-atomic world, in terms of quantum field, is closely paralleled by that of the Eastern mystic who interprets his or her experience of the world in terms of an ultimate underlying reality", says Fritz Capra.[34]

Empty space-void is the seat of violent physics

(John F. Wheeler)

What is an empty space in quantum physics? It is space where pairs of particles, pairs of photons or gravitons continually appear. The two particles in a pair start out together, then move apart. After an interval of time too short to imagine they come together again and annihilate one another a short but eventful life. Quantum mechanics tells us this is happening all the time, everywhere in the "vacuum" of space.[35]

The universe did not dwindle to zero

Uncertainty principle, the foundation of today's quantum mechanics means that so-called empty space teams with particles and anti-particles. These particles add up to an enormous amount of energy — an infinite amount.

The presence of this energy causes curvature of space and time, according to Relativity theory. "Combining the two, we must conclude that the entire universe ought to be curled up into a small ball. Curiously, this has not happened. Both the general relativity and quantum mechanics are exceptionally good theories which has been put into extensive practice. Nevertheless, put together they yield infinites and not zeroes".[36]

"Real" out of "Virtual" Particles

The density of the matter in a neutron star, in grams per cubic centimetre, would be 10/14 - that is, 1 followed by 14 zeroes, or one hundred thousand billion. Even an object this dense would not be a black hole, for light could still escape from its surface into the universe at large. Making a black hole from a dead star would require, crushing even neutrons out of existence. The gravitational field at the even horizon of black hole is strong enough to do an astounding thing. Stephen Hawking predicted that it could change the "virtual" particles with negative energy to "real" particles with positive energy.[37]

Empty space or void is an impossibility

The upshot of the uncertainty principle is that no field can ever measure zero. Since the equation of the uncertainty principle precludes anything from becoming a zero, zero would be a very precise measurement of both the value of the field and its rate of change, but the uncertainty principle would not allow that. We do not have empty space unless all fields are exactly zero; *therefore no zero—no empty space.*

In 1970 Stephen Hawking mathematically postulated that even black holes can rotate and radiate energy. This radiation known as Hawking radiation fobs the black hole of energy. When something has less energy, it automatically has less mass.

The black hole therefore should have ended in zero mass. Uncertainity Principle shows that that, however, never happens. At near infinite density but near zero mass as Hawking thinks, the black hole bursts out to new creation. Nobody is sure how it ends. Hawking guesses according to uncertainty principle that the little black hole disappears in a huge final puff of particle emission, like millions of hydrogen bombs exploding. The idea that a black hole could get smaller and finally explode was declared by Hawkin's contemporaries as "absolute rubbish" but many eminent physicists welcomed this as their greatest discovery combining successful quantum mechanics with Relativity.[38]

Einstein theorized about something called the cosmological constant, which would balance gravity and prevent the universe from changing in size. He later called it "the greatest blunder of my life". The term today has come to have a related but slightly different meaning. The cosmological constant, as scientists now use the term, is a number which tells us how dense is the energy in the vacuum. Common sense says there should not be any energy in the empty space at all, but the uncertainty principle shows that "empty" space is not empty. It seethes with infinite energy. The cosmological constant (the energy density of the vacuum) therefore ought to be enormous and the general relativity theory tells us this enormous mass energy should be curling up the universe, into a zero volume. Yet paradoxically what happens is quite the contrary. The value of the cosmological constant is observed to be near-zero. How can the cosmological constant possibly be near zero, as we observe, when theory tells us it should be enormous, almost infinite? Sidney Coleman, who shares Hawking's enthusiasm for worm holes, puts it, "Zero is a suspicious number".[39]

The Chinese concept of *Ch'i* bears resemblance to the concept of the quantum field in modern physics. Like the quantum field, *Ch'i* is conceived as a tenuous and non-perceptible form of matter which is present throughout space, and can condense into solid material objects. In the words of Chang Tsai :

"When the *Ch'i* condenses, its visibility becomes apparent so that there are then the shapes (of individual things). When it disperses, its visibility is no longer apparent and there are no shapes. At the time of its condensation, can one say otherwise than that this is but temporary? But at the time of its deciphering, can one hastily say that it is then non-existent?"[40]

Emptiness condenses into forms and again disperses into void. As Chang Tai says again, "The great Void cannot but consist of *Ch'i*; this *Ch'i* cannot but condense to form all things and these things cannot but become dispersed so as to form (once more) the Great void".[41]

The field exists always and everywhere. It can never be removed. It is the carrier of all material phenomena. It is the 'void' out of which the proton creates the pi-mesons. Fading of particles and their rebirth in new form are merely forms of motion of the field, a discovery of modern physics, demonstrated by Feynmann's diagram.

Particle physics show that virtual particles can come into being spontaneously out of the void and vanish again into the void, without any nucleon or other strongly interacting particle being present. Physicist Wheeler says that "Empty space is the place of most violent physics".[42]

Like the *śūnyatā* or void, the 'physical vacuum' — as it is called in field theory — is not a state of mere nothingness, but contains the potentiality for all forms of physical entities. The Buddhist *sūtra* says, "*Form is emptiness and emptiness is indeed form*".

"The vacuum is truly a 'living void', pulsating in endless rhythms of creation and destruction".

Upaniṣads discovered some thousand years ago that Brahman, the massless entity beyond space-time which is known as *neti neti* (not this, not this) or *nothing* or *śūnya* in space-time contains potentially the infinity of existence-knowledge-bliss.

REFERENCES

1. *The Gospel of Sri Ramakrishna*, p. 870 (translation by Swami Nikhilananda, New York).
2. *Ibid.*, p. 672.
3. *Ibid.*, p. 947.

4. *Ibid.*, p. 947.
5. *A Study of History*, Arnold J. Toynbee, I-IV, Abridgement by D. C. Somerville, p. 390.
6. *Buddhism*, Punjab University. Patiala 1969. p. 78.
7. *2500 Years of Buddhism*, editor P.V. Bapat, Ministry of Information, Govt. of India, p. 106.
8. *A History of Indian Philosophy*, Cambridge University Press 1951, Vol. I, p. 140.
9. *Ibid.*, p. 140.
10. *Ibid.*, p. 147.
11. *Ibid.*, p. 149.
12. *Tao of Physics*, Fritz Capra, 1975.
13. *Ibid.*, p. 104.
14. *Ibid.*, p. 104.
15. *Ibid.*, p. 105.
16. *Ibid.*, p. 110.
17. 2500 Years of Buddhism, p. 107.
18. Christmas Humphreys, *Buddhism*, Pelican Book, 1951, pp. 107, 108, 109, 149.
19. *Mahāyāna Sūtrālaṅkāra of Asaṅga*, Dr Yogeswar Shastri, Satguru Publication, Delhi, 1989, pp. 11, 127, 138-139.
20. *Ibid.*, Introduction.
21. *The Brahma Sūtra* : Dr. S. Radhakrishnan, George Allen and Unwin, 1960 p. 378.
22. *Ibid.*, p. 63.
23. *Ibid.*, p. 378-79.
24. *Indian Philosophy*, Dasgupta, p. 257.
25. *Ibid*, Vol. 6, p. 211.
26. *Ibid*, Vol. 8, p. 11.
27. *Ibid*, Vol. 3, p. 28.
28. *Ibid*, Vol. 2, p. 411.
29. *Ibid*, Vol. 2, p. 415.
30. *Ibid*, Vol. 2, pp. 416-17.
31. *Tao of Physics* : Fritz Capra, p. 220.
32. *Stephen Hawking* by Michael White & John Gribbin (Vikina) 1992, p. 179.
33. *Tao of Physics*, Fritz of Capra, pp. 220-21.
34. *Ibid.*, p. 221.
35. *Stephen Hawking*, Kitti Ferguson, Bantam, 1992, pp. 77-78.
36. *Ibid.*, p. 24.
37. *Stephen Hawking*, by M. White & J. Gribbin pp. 76-77.

38. *Stephen Hawking*, by K. Ferguson p. 82
39. *Ibid*., pp. 146-47.
40. *Tao of Physics*, p. 225.
41. *Ibid*., p. 225.
42. *Ibid*., p. 234.

Technology of Using Śūnya in India

R.C. GUPTA

There is no doubt that very many ideas and concepts are involved when we talk of *śūnya* in Indian context. It is understood that the earliest literary uses of the word *śūnya* were related to its dictionary meanings namely void, empty, vacant, wanting etc. A couplet from the ancient *Cāṇakya-nīti* reads

avidyaṃ jīvanaṃ śūnyaṃ, diśaḥ śūnyās tv abāndhavāḥ /
mūrkhasya hṛdayaṃ śūnyaṃ, sarva-śūnyā daridratā // (IV, 14)

'Life is *śūnya* or void without learning; surroundings are *śūnya* without company; a fool's heart is void; all is void in poverty.'

Here we clearly have the use of *śūnya* to indicate the sense of voidness, emptiness, blankness or some sort of absence of something. Formation of compounds like *śūnyaśālā* and *śūnyavājin* ('a horse without rider') was common. Interestingly, and befitting to the saying '*sarva-śūnyā daridratā*', the compound *śūnya-hara* (remover of wantingness) is taken to mean 'gold' by lexicographers.

Derived from the basic sense of voidness or absence, the word *śūnya* was used later on in somewhat technical sense in philosophy and other speculative systems like mysticism, grammar and linguistics. For instance, the Buddhist doctrine of *śūnyavāda* ('zeroism') is related to the absence of any basic cause or principle or spirit either human or supreme (the saying '*sarvaṃ śūnyaṃ*' is attributed to Nāgārjuna of about A.D. 150). According to *Sarvadarśana Saṅgraha*, *śūnyatā* or *śūnyatva* is the illusory nature of all worldly phenomena.

The above non-mathematical (literary and speculative) use of *śūnya* (or zero) may be said to form the first or the earliest phase of the technology of using zero in India. Subsequently there were semi-mathematical and mathematical uses of the concept of zero. Each of these may be put into two categories. Thus in all we have the following six categories in which the use of the *śūnya* or zero concepts may be broadly conceived:

(i). Purely literary
(ii). Speculative
(iii). Referencial
(iv). Place-value numeration
(v). Computable number
(vi). Non-scalar or higher mathemtical entity.

This scheme of characterizing the role of zero, although not perfect, may serve as a guidance in discussing the history and chronology of evolution of the various concepts and ideas related to *śūnya* in its multidimensional aspect.

Before giving details, it may be pointed out that a large number of words synonymous to *śūnya* have been used in ancient and medieval Indian works. These are listed in the Appendix at the end. Of course, the use of a large number of synonyms is a usual feature of Sanskrit language. The important point to note is that the use of a large number of terms of various nature easily leads to multiple interpretations. For instance, the use of the word *ananta* (infinite) as a synonym of zero may impart the small circle or dot (*śūnya-bindu*) the status of 'infinity', or zero may be called *ananta* (endless) because the (small) circle has no end. According to Apte's Dictionary, *śūnya* is a name of Brahma.

It is well-known that as a word-numeral, the term *pūrṇa* was used for zero. In this context we have the following Upaniṣadic verse:

pūrṇam adaḥ pūrṇam idaṃ pūrṇat pūrṇam udacyate ।
pūrṇasya pūrṇam ādāya pūrṇam evāvaśiṣyate ।।

'That is *pūrṇa*, this is *pūrṇa*; *pūrṇa* has been evolved from *pūrṇa*. Taking the *pūrṇa* from the *pūrṇa*, what remains is (still) *pūrṇa*.'

Philosophically, *pūrṇa* ('perfect') here may be taken to mean 'infinite fullness'. Mathematically the second half implies:

$$\text{pūrṇa} - \text{pūrṇa} = \text{pūrṇa}$$

or

$$x - x = x$$

which is always satisfied by $x = 0$, but can also be satisfied when x is infinity in many limiting cases.

Speculative claims have been made for grammatical and linguistic zero comparable to use of zero in mathematical sense. In Indo-European philology, we have the 'zero-grade form' which is formed by dropping an 'e', for example in the following cases of roots

√*yeug* ("to join") → *yug*; cf. *yuga*, yoke etc.

and √*gher* ("to enclose") → *ghr*; cf. *gṛha*, garden etc.

Pāṇini's use of his grammatical or linguistic zero ('*lopa*') is used as a maker of an empty ('*śūnya*') non-occupied space or position. His relevant basic *sūtra* is

adarśanaṃ lopaḥ (*Aṣṭādhyāyī*, I. 1.60)

'Non-appearance (of a sound or morpheme) is *lopa*.'

That the Pāṇinian '*lopa*' stood for a sort of zero is also indicated by the fact that Pūjyapāda (A.D. 450 nearly) in his *Jainendra Vyākaraṇa* replaced '*lopa*' in the *sūtra* VIII.4.64 by '*khaṃ*' which was a common term for zero.

In fact, as far as speculative thoughts (including *yoga*, *tantra*, mysticism etc.) are concerned we do come across many diverse and divergent meanings and views. *Śūnya* and *pūrṇa* are taken "as modalities of consciousness in the human quest for beauty and the soul". Often the state of egolessness is compared with *śūnya* and is important for a yogi for the ascent of the soul (zero level) to Brahma (infinite fullness).

Speaking of the third category, we might say that *śūnya* or void of the empty space may be the point of reference from which all existence (at least in the present cycle) emerged (cf. the big bang theory of astronomers). Scientifically this category consists of those ideas and concepts in which zero is used as an indicator of the point of reckoning, measuring, or graduation. Zero hour is the scheduled time for a specific action or operation. There was a day whose Julian Day Number was zero. In astronomical calculations we often have different epochs when the *ahargaṇa* is zero. For the calibration of a scale measuring any physical quantity we use a zero mark, e.g. for indicating zero temperature. For measuring distances along any line or curve (e.g. a road) we use and mark a reference point and call it our zero. It is said that the ancient Egyptians used the hieroglyphic symbol *nfr* to mark the ground level reference line at the Medium Pyramid belonging to the Old Kingdom. The mean sea-level is the zero to measure heights above (+ ve) and depths below (–ve).

Grown from the original idea of emptiness, the next category deals with the concept of *śūnya* or zero as a symbol which denotes the absence of a denominational term in any positional numerical notation. The sign or indicator for zero may be merely a 'vacant space' or it may be some concrete symbol to avoid any visual confusion. Categories three and four may be considered to be semi-mathematical because in these the technology or process of using zero is non-mathematical although the evolved system is mathematical.

The present decimal place-value system, which is now used all over the civilized world, is a very clear and ideal example of a positional numerical notation with a concrete zero symbol. The system is the greatest scientific gift of India to the world. It must be noted that in a fully developed place-value system of numeration, the zero must be able to play all the following three roles successfully :

(1) medial or internal, which is the classical role of a blank space, e.g. as in 205, 2005 etc.
(2) final or terminal, which is more stringent role, e.g. in 250, 2500 etc.
(3) initial, which is rather a superfluous role ordinarily, e.g. 025 = 0025 = 25 in value. But in computer age this role is also important.

Moreover, for a perfect and ideal system to base *b* there must be only one set of *b*–1 non-zero independent numerical signs and the same zero symbol must play all the above three roles (pre-Indian examples often lack these ideal conditions).

As far as the date of invention of the decimal place-value system with zero is concerned, it is almost impossible to know. Such difficulty is partially created by the practice of attributing divine origin to all arts and sciences in ancient India. But a symbol for zero (*śūnya*) was used by Piṅgala who is placed about 200 B.C.

In the next category is placed the discussion of zero as a number (or *saṁkhyā*). According to some scholars the word *kṣudra* occurring in the *Atharva-Veda, Kāṇḍa* XIX, *Sūkta* 22-23, is taken to stand for zero along with the words *ṛca* and *anṛca* which are taken to indicate positive and negative numbers. All ancient Indian mathematical or astronomical works imply the use of calculations with zero as such or via the system of word-numerals (*bhūta-saṃkhyā*) and letter-numerals. But it was

Brahmagupta (A.D. 628) who formally dealt with the mathematics of zero in detail. He laid down the rules which the computable number zero obeyed when subjected to mathematical operations and laws. In his *Brāhmasphuṭa-siddhānta* (XVIII, 30) he defines zero by :

$$n + (-n) = 0$$

He clearly stated equivalents of

$$x \pm 0 = x; \quad 0 \pm 0 = 0 \text{ and } x \times 0 = 0 = 0 \times 0$$

The first two of these six equations show the trifling (*kṣudra* or *tuccha*) nature of zero. Brahmagupta seems to have confused in his exposition when division by zero was involved. Śrīdhara (about A.D. 750) also gave a good exposition of zero in his *Triśatikā*, but it is unfortunate that the *Śūnyatatva* portion in his bigger *Pāṭīgaṇita* is lost.

In the above fifth category, the discussion of zero was confined to the system of real (or ordinary) numbers. But in modern mathematics, we have more general or comprehensive systems of numbers and quantities such as complex and hyper-complex numbers, vectors and other non-scalar quantities. The discussion of zero in all such higher mathematical systems is supposed to be dealt in the sixth category. But here we mention about this only briefly. The zero element in these systems commonly follows the rule : $S + 0 = S$. But the zero vector or zero matrix is quite different from zero as a real number. Even two zero matrices may not be same. And so on.

Finally, it may be reminded that the above classification into six categories is for convenience. Discussion of ideas and concepts behind *śūnya* or zero from ancient as well as modern topics may need more or special categories.

Anyway, we find that the concept of *śūnya* or zero provoked ideas which engulfed almost all fields of human endeavour. It played significant role in arts, sciences and other fields not only in India but in the whole world. The story of the spread and triumph of the Indian decimal place-value system with zero is indeed marvellous because it revolutionised the scientific development all over the globe. It was also a cultural victory for India brought not by sword but by merit of the case.

APPENDIX

Some Words, Letters and Word-Numerals denoting Zero

ananta
antarikṣa
abhra
ambara
ambuda
kha (= *chidra*)
khaṃ (= *ākāśa*)
gagana
ṅa (in *Ekākṣarakośa*)
chidra
jalada
jaladhara
jaladharapatha
jha (in *Ekākṣarakośa*)
jña (in *Kaṭapayādinyāya*)
ṭha (in *Ekākṣarakośa*)
tuccha (= *kṣudra*)?
div or *diva*
dyo or *dyau*
na (in *Kaṭapayādinyāya*)
nagna
nabha
payoda
payodhara
asat
ākāśa
kṣa (in Tirthaji's *Vedic mathematics*)
kṣudra? (in *Atharva-Veda*)
puṣkara
pūrṇa
pṛthu
pṛthula
bindu
megha
randhra
vārīya
vindu
viyat
viṣṇupada
viṣṇupāda
vihāyas
vṛhat
vyoma
śunya
śūnya
suravartma?

Vedic Numerical System including Śūnya

S.A.S. SARMA

Numeration, mathematics and astronomy were but natural and necessary concomitance with the civilization of the times in India. An idea of the advances reached during Vedic times in different disciplines can be visualised if one is aware of the several technical and semi-technical professions practised during those days as mentioned in different contexts in Vedic literature. Among these professions are those of the carpenter (*takṣan, ṚV* 9.112.1), smith (*karmakāra, ṚV*. 10.72.2), one who smelted metal in fire (*dhūmātṛ, AV* 3.5.6), ploughman (*kṛṣṭi, ṚV*. 1.177.1; 4.17.5), barber (*vaptṛ, ṚV*. 10.142.4; *nāpita, ŚB*, 3.1.2.2), fisherman (*puñjiṣṭha, YS-VS*. 16.27; *TS* 4.5.4.2), boatman (*nāvaja, ŚB*. 2.3.3.5), jeweller (*maṇikāra, TB* 3.4.31.1), basket-maker (*biḍālakāra, YV-VS*. 30.8; *TB*. 3.4.5.1), washer-man (*malaga, AV*. 12.3.21), dyer (*rajayitṛ, YV-VS*. 30.12, *TB*. 3.4.7.1), rope-maker (*rajjusarja, YV-VS*. 30.7; *TB*. 3.4.3.1) and weaver (*vayitṛ, Pañcaviṃśabrāhmaṇa* 1.8.9). Two further professionals were the mathematician (*gaṇaka, YV-VS*. 30.20; *TB*. 3.4.15.1) and astronomer (*nakṣatra-darśa, YV-VS*. 130.10; *TB*. 3.4.4.1). It would be no wonder that a society which had acquired proficiency in the above-said several trades had developed their numeral systems as well for facilitating the proper functioning and development of their professions. In fact, it might be seen from the discussion below that the decimal numeration adumbrated in the Vedic literature is as perfect in its fundamentals as the numeral system of modern times.

The importance of the decimal system is very clear from the words of the great mathematician Laplace who observes: "The idea of expressing all quantities by nine figures whereby is imparted to them both an absolute value and one by position is so simple that this very simplicity is the reason for our not being sufficiently aware how much admiration it deserves". This system imparts to the figures (digits) an absolute numerical value and also a positional value which latter increases their value ten times by being shifted by one place to the left. Thus, the digits: 1, 2, 3, 4, 5, 6, 7, 8 and 9, coupled with the figure '0' which stands for zero or '*śūnya*' ('nothing, empty'), while expressing just their individual values when standing alone, can express also any quantity of any magnitude by their repeated use in the same number just by shifting their places, as needed. The '*śūnya*' or the zero bears a major role in this system. This system was in vogue from the Vedic times as it could be seen in the passages of the *Ṛgveda* where it groups the gods into three (*ṚV*. 1.105.5), the sages were seven (*ṚV*. 4.42.8) and so were the seas (*ṚV*. 8.40.5). The Marut-gods were 180 or three times sixty (*ṚV*. 8.96.8), god Śyāvā gifted cows numbering 210 or three times seventy (*ṚV*. 8.19.37). The followers of Indra were 21 or three times seven (*ṚV*. 1.133.6) and the number of horses prayed for was thrice seven times seventy or 3 x 7 x 70 (*ṚV*. 8.46.26).

In order to illustrate that 'things are spoken in accordance with their states', Vasumitra, the Buddhist teacher says, "When the clay counting-piece is placed in the place of units, it is denominated 'one', when placed in the place of hundreds, it is denominated 'hundred' and in the place of thousands it is denominated a 'thousand'". This clearly shows the use of decimal place-value system during his period.[1] In the ancient Jaina works too one can see similar process mentioned. The popularity of the decimal system can be noticed from the following line of *Vyāsabhāṣya* on the *Yogasūtra* III.13:[2] '*yathaikā rekhā śatasthāne śataṃ, daśasthāne daśa, ekā caikasthāne*', i.e. "as the figure 1 is hundred in the place of hundred, ten in the place of ten and one in the place of one'.

In Vedic literature are found, besides the primary numbers, one to nine, expressed by the terms, *eka, dvi, tri, catur, pañca, ṣaṭ, sapta, aṣṭa* and *nava*, the decuple terms from ten to ninety, expressed by *daśa, viṃśati, trimśat, catvāriṃśat, pañcāśat, ṣaṣṭi, saptati, aśīti* and *navati* and these sequentially multiplied by ten, taking terms from 100 to 10 to the power of 12, the terms being, *śata* (10^2), *sahasra* (10^3), *ayuta* (10^4), *niyuta* (10^5), *prayuta* (10^6), *arbuda* (10^7), *nyarbuda* (10^8), *samudra* (10^9), *madhya* (10^{10}), *anta* (10^{11}) and *parārdha* (10^{12}).

In the matter of the arrangement of decuples in compound number-names, the practice generally followed in Vedic literature was to put the term of higher denomination, first, except in the case of the two lowest denominations, where the reverse method was followed. See, for example, *sapta śatāni viṃśatiḥ* (seven hundreds and twenty, *ṚV.* 1.164.11), *sahasrāṇi śata daśa* (thousands hundred and ten, *ṚV.* 2.1.8) and *ṣaṣṭiṃ sahasra navatiṃ nava* (sixty thousands, ninety and nine, 60,099, *ṚV.* 1.53.9).

It is also to be noted that in expressing a number greater than 10^3 (*sahasra*), it was more usual to follow a centesimal scale. Thus we could find *ṣaṣṭiṃ sahasrāṇi* or 60 x 10^3 (*ṚV.* 1.53.9; *ŚB.* X.2.1.11, XIII.4.16), *pañcāśat sahasram* or 50 x 10^3 (*ṚV.* IV.6.13), *dvā-saptati sahasrāṇi* or 72 x 10^3 (*Bṛhadāraṇyakopaniṣad* I.1.19) etc.

As pointed out by Pāṇini in his *Aṣṭādhyāyī* (V.I.59),[3] all the number-names except the numbers up to ten are formed on the multiplicative principle. The compound name for a number below hundred is formed with the principle of addition, such as the usage in *ekādaśa* (11), *ekaviṃśati* (21) etc. The subtraction also could be seen in the following usage such as *ekān-na-viṃśati* (or *nava-daśa*; 20—1), *ekān-na-triṃśat* (or *nava-viṃśati*; 30—1), *ekān-na-śata* (or *nava-navati*; 100—1) etc.

In the *TS* (VII.2.11)[4] we could see the usage of subtraction and addition which are used alternatively thus: *ekān-na-viṃśati, nava-viṃśati, ekān-na-catvāriṃśat, nava-catvāriṃśat* and so on. But later the terms like *nava-daśa*, *nava-viṃśati* became obsolete and the principle of substraction was used to denote such numbers. Here too the prefix '*ekānna*' changed to '*ekona*' and the usage became '*ekona-viṃśati*' and '*ekona-triṃśat*'. It is also noteworthy that multiplication also is indicated in the passages of *Ṛgveda* (*ṚV.* VIII.9.68) where it uses the number 180 as three sixties; 210 as three seventies (*ṚV.* VIII.19.37), 21 as three sevens (*ṚV.* I.133.6, 191.12) and 1470 as three seven seventies (*ṚV.* VIII.46.26).

The sequential mention of *yugma* (even) and *ayugma* (odd) is also seen in Vedic texts. The series or progression of *ayugma* and *yugma* numbers occur in the well known *camaka sūkta* (*TS.* 4.7.11) where the odd numbers refer to the *chandas* of the gods and the even numbers refer to that of men.

ekā ca me tisraś ca me pañca ca me sapta ca me nava ca me ekādaśa ca me trayodaśa ca me pañcadaśa ca me saptadaśa ca me navadaśa ca me ekaviṃśatiś ca me trayoviṃśatiś ca me pañcaviṃśatiś ca me saptaviṃśatiś ca me navaviṃśatiś ca ma ekatriṃśac ca me trayastriṃśac ca me catasraś ca me 'ṣṭau ca me dvādaśa ca me ṣoḍaśa ca me viṃśatiś ca me caturviṃśatiś ca me 'ṣṭāviṃśatiś ca me dvātriṃśac ca me ṣaṭtriṃśac ca me catvāriṃśac ca me catuś catvāriṃśac ca me 'ṣṭācatvāriṃśac ca me vājaś ca prasavaś cāpijaś ca kratuś ca suvaś ca mūrdhā ca vyaśniyaś cāntyāyanaś cāntyaś ca bhauvanaś ca bhuvanaś cādhipatiś ca.[5]

The enumeration of numbers in the ascending order and descending order also was in use. See, for instance, the ascending order occurring in the *Śatarudrīya*, in which the first set of eleven refers to the group of Vasus, the second to the Rudras and the third to the Ādityas. (*TB.* 3.11.2).

devā deveṣu śrayadhvam. prathamā dvitīyeṣu śraydhvam. dvitīyās tṛtīyeṣu śrayadhvam. tṛtīyāś caturtheṣu śrayadhvam. caturthāḥ pañcameṣu śrayadhvam. pañcamāḥ ṣaṣṭheṣu śrayadhvam. ṣaṣṭhāḥ saptameṣu śrayadhvam. saptamā aṣṭameṣu śrayadhvam. aṣṭamā navameṣu śrayadhvam. navamā daśameṣu śrayadhvam. daśamā ekādaśeṣu śrayadhvam. ekādaśā dvādaśeṣu śrayadhvam. dvādaśās trayodaśeṣu śrayadhvam. trayodaśāś caturdaśeṣu śrayadhvam. caturdaśāḥ pañcadaśeṣu śrayadhvam. pañcadaśāḥ ṣoḍaśeṣu śrayadhvam. ṣoḍaśās saptadaśeṣu śrayadhvam. saptadaśā aṣṭādaśeṣu śrayadhvam. aṣṭādaśā ekān-na-viṃśeṣu śrayadhvam. ekān-na-viṃśā viṃśeṣu śrayadhvam. viṃśā ekaviṃśeṣu śrayadhvam. ekaviṃśā dvāviṃśeṣu śrayadhvam. dvāviṃśās trayoviṃśeṣu śrayadhvam. trayoviṃśaś caturviṃśeṣu śrayadhvam caturviṃśaḥ pañeaviṃ śeṣu śrayadhvem pañcaviṃśāḥ ṣaḍviṃśeṣu śrayadhvam. ṣaḍviṃśāḥ saptaviṃśeṣu śrayadhvam. saptaviṃśā aṣṭāviṃśeṣu śrayadhvam. aṣṭāviṃśā ekān-na-triṃśeṣu śrayadhvam. ekān-na-triṃśās triṃśeṣu śrayadhvam. triṃśā ekatriṃśeṣu śrayadhvam. ekatriṃśā dvātriṃśeṣu śrayadhvam. dvātriṃśās trayastriṃśeṣu śrayadhvam. devās trir ekādaśās tris trayastriṃśāḥ.

The enumeration of numbers in the *ātmarakṣā* (*TB.* 2.3.11) forms an example of the descending order of numbers.

brahmātmanvad asṛjata. tad akāmayata. samātmanā padyeyeti. ātmann ātmann ity āmantrayata. tasmai daśamaṃ hūtaḥ pratyaśṛṇot. sa daśahūto 'bhavat. daśahūto ha vai nāmaiṣaḥ. taṃ vā etaṃ daśahūtaṃ santam. daśahotery ācakṣate parokṣeṇa. parokṣapriyā iva hi devāḥ.
ātmann ātmann ity āmantrayata. tasmai ṣaṣṭhaṃ hūtaḥ pratyaśṛṇot. sa ṣaḍḍhūto 'bhavat. ṣaḍḍhūto ha vai nāmaiṣaḥ. taṃ vā etaṃ ṣaḍḍhūtaṃ santam. ṣaḍḍhotety acakṣate parokṣeṇa. parokṣapriyā iva hi devāḥ.
ātmann ātmann ity āmantrayata. tasmai pañcamaṃ hūtaḥ pratyaśṛṇot. sa pañcahūto 'bhavat. pañcahūto ha vai nāmaiṣaḥ. taṃ vā etaṃ pañcahūtaṃ santam. pañcahotety ācakṣate parokṣeṇa. parokṣapriyā iva hi devāḥ.
ātmann ātmann ity āmantrayata. tasami caturthaṃ hūtaḥ pratyaśṛṇot. sa caturhūto 'bhavat. caturhūto ha vai nāmaiṣaḥ. taṃ vā etaṃ caturhūtaṃ santam. caturhotety ācakṣate parokṣeṇa. parokṣapriyā iva hi devāḥ.

An instance of arithmetical progression occurs in the *TS* (VI.2.12-17) in the context of successive numbers of oblation to be made to Prajāpati in the consecutive hoof-marks of the horse of the Aśvamedha sacrifice as it returns to the sacrificial field after travelling the entire land for a period of one year.

(1) 1, 3, 5, . . . 19 . . . 29, . . . 99

ekasmai svāhā tribhyaḥ svāhā pañcabhyaḥ svāhā saptabhyaḥ svāhā navabhyaḥ svāhaikādaśabhyaḥ svāhā trayodaśabhyaḥ svāhā pañcadaśabhyaḥ svāhā saptadaśabhyaḥ svāhaikān-na-viṃśatyai svāhā navaviṃśatyai svāhaikān-na-catvāriṃśate svāhā navacatvāriṃśate svāhaikān-na-ṣaṣṭyai svāhā navaṣaṣṭyai svāhaikān-nāśītyai svāhā navāśītyai svāhaikān-na-śatāya svāhā śatāya svāhā sarvasmai svāhā //TS. VI.2.12//

(2) 2, 4, 6, . . . 100

dvābhyāṃ svāhā caturbhyaḥ svāhā ṣaḍbhyaḥ svāhā 'ṣṭābhyaḥ svāhā daśabhyaḥ svāhā dvādaśabhyaḥ svāhā caturdaśabhyaḥ svāhā ṣoḍaśabhyaḥ svāhā 'ṣṭādaśabhyaḥ svāhā viṃśatyai svāhā 'ṣṭānavatyai svāhā śatāya svāhā sarvasmai svāhā // 13 //

(3) 3, 5, 7, ...100

tribhyaḥ svāhā pañcabhyaḥ svāhā saptabhyaḥ svāhā navabhyaḥ svāhaikādaśabhyaḥ svāhā trayodaśabhyaḥ svāhā pañcadaśabhyaḥ svāhā saptadaśabhyaḥ svāhaikān-na-viṃśatyai svāhā navaviṃśatyai svāhaikān-na-catvāriṃśate svāhā navacatvāriṃśate svāhaikān-na-ṣaṣṭyai svāhā navaṣaṣṭyai svāhaikān-nāśītyai svāhā navāśītyai svāhaikān-na-śatāya svāhā śatāya svāhā sarvasmai svāhā//14//

(4) 4, 8, 12, . . . 100

caturbhyaḥ svāhā 'ṣṭābhyaḥ svāhā dvādaśabhyaḥ svāhā ṣoḍaśabhyaḥ svāhā viṃśatyai svāhā ṣaṇṇavatyai svāhā śatāya svāhā sarvasmai svāhā//15//

(5) 5, 10, 15 . . . 100

pañcabhyaḥ svāhā daśabhyaḥ svāhā pañcadaśabhyaḥ svāhā viṃśatyai svāhā pañcanavatyai svāhā śatāya svāhā sarvasmai svāhā//16//

(6) 10, 20, 30 . . . 100

daśabhyaḥ svāhā viṃśatyai svāhā triṃśate svāhā catvāriṃśate svāhā pañcāśate svāhā ṣaṣṭyai svāhā saptatyai svāhā 'śītyai svāhā navatyai svāhā śatāya svāhā sarvasmai svāhā //TS. VI.2.17//

The thirty fundamental terms of numbers are enumerated in a series in the *Śivasaṅkalpopaniṣat.*[6] It reads thus:

ekā ca daśa śataṃ ca sahastraṃ cāyutaṃ ca
niyutaṃ ca prayutaṃ cārbudaṃ ca nyarbhudaṃ ca/
samudraś ca madhyaṃ cāntaś ca
parārdhaś ca tan me manaḥ śivasaṅkalpam astu.//[7]

In continuation the *TS* (VII.2.20)[8] enumerates the number up to *loka* (10^{19}) namely, *uṣas* ($=10^{13}$), *vyuṣṭi* ($=10^{14}$), *udeṣyat* ($=10^{15}$), *udyat* ($=10^{16}$), *udita* ($=10^{17}$), *suvarga* ($=10^{18}$) and *loka* ($=10^{19}$).

The enumeration of the exact strength of Rāma's army in the *Rāmāyaṇa*[9] (6.28.33-41) is noteworthy here since it gives a beautiful description of the number terms. It says:

śataṃ śatasahasrāṇāṃ koṭim āhur manīṣiṇaḥ/
śataṃ koṭisahasrāṇāṃ śaṅkur ityabhidhīyate//
śataṃ śaṅkusahasrāṇāṃ mahāśaṅkur iti smṛtaḥ/
mahāśaṅkusahasrāṇāṃ śataṃ vṛndam ihocyate//
śataṃ vṛndasahasrāṇāṃ mahāvṛndam iti smṛtam/
mahāvṛndasahasrāṇāṃ śataṃ padmam ihocyate//
śataṃ padmasahasrāṇāṃ mahāpadmam iti smṛtam/
mahāpadmasahasrāṇāṃ śataṃ kharvam ihocyate//
śataṃ kharvasahasrāṇāṃ mahākharvam iti smṛtam/
mahākharvasahasrāṇāṃ samudram abhidhīyate//
śataṃ samudrasāsram ogha ityabhidhīyate/
śatam oghasahasrāṇāṃ mahaugha iti viśrutaḥ//
evaṃ koṭisahasreṇa śaṅkūnāṃ ca śatena ca/
mahāśaṅkusahasreṇa tathā vṛndaśatena ca//
mahāvṛndasahasreṇa tathā padmaśatena ca/
mahāpadmasahasreṇa tathā kharvaśatena ca//
samudreṇa ca tenaiva mahaughena tathaiva ca/
eṣa koṭimahaughena samudrasadṛśena ca/
vibhīṣaṇena vīreṇa sacivaiḥ parivāritaḥ//

This means the following enumeration :

100 *lakh* = 1 *koṭi* (crore) = 10^7
10^5 *koṭi* =1 *śaṅku* = 10^{12}
10^5 *śaṅku* = 1 *mahāśaṅku* = 10^{17}
10^5 *mahāśaṅku* = 1 *vṛnda* = 10^{22}
10^5 *vṛnda* = 1 *mahāvṛnda* = 10^{27}
10^5 *mahāvṛnda* = 1 *padma* = 10^{32}
10^5 *padma* = 1 *mahāpadma* = 10^{37}
10^5 *mahāpadma* = 1 *kharva* = 10^{42}
10^{10} *kharva* = 1 *mahākharva* = 10^{47}
10^3 *mahākharva* = 1 *samudra* = 10^{50}
10^5 *samudra* =1 *mahaugha* = 10^{55}

Rāma's army is thus stated as 1000 *koṭi* + 100 *śaṅku* + 1000 *mahāśaṅku* + 100 *vṛnda* + 1000 *mahāvṛnda* + 100 *padma* + 1000 *mahāpadma* + 100 *kharva* + 100 *samudra* + 100 *mahaugha* + 1 *koṭi mahaugha* + 1 vibhīṣaṇa and his four ministers, which is equal to $10^{10} + 10^{14} + 10^{20} + 10^{24} + 10^{30} + 10^{34} + 10^{40} + 10^{44} + 10^{52} + 10^{57} + 10^{62} + 5$.

In the Buddhistic work *Lalitavistara*[10] of the first century B.C., Bodhi Sattva enumerates to a mathematician Arjuna the system of numerals in multiples of 100, starting from *koṭi* (10^7) up to 10^{53}.

The geometrical progression of numbers like 12, 24, 48, 96, 192, . . . etc. occurs in the *Pañcaviṃśabrāhmaṇa* (XVIII.3) while describing the list of sacrificial gifts.[11]

In the Vedic period itself, the numbers three and its multiples and seven and its multiples got a prominent place among the several numbers which are considered as sacred. The earliest division of the world into three, viz. heaven, sky and earth, itself could be a good example. The seven regions and seven places of the earth are mentioned in the *Ṛgveda* (IX.114.3, I.22.16). The number seven occurs also in connection with Agni, as he had seven wives in the form of seven flames, the sevenfold song for Agni, the seven horses of the Sun, the group of seven ṛṣis, the seven *hotṛs*, the seven layers of twenty-one bricks for the fire-altar and the seven steps or '*saptapadī*'[12] performed in the marriage ritual are other examples in this regard. The numbers nine, twelve, eighteen are also well known for their auspiciousness.[13]

One can also see the usage of word-numerals in the Vedic texts even though this system was widely used in the discipline of astronomy and mathematics. Metres like *gāyatrī* and *jagatī* are used as word-numerals to denote 24 and 48 respectively in the texts of *Kātyāyana-śrautasūtra* and *Lāṭyāyana-śrautasūtra.*[14] In Vedic literature the words *kalā* (ṚV. 8.47.17), *kuṣṭha (YV. Maitrāyaṇīyasaṃhitā* 8.7.7), *śapha* (ṚV. 8.47.17) and *pāda* (ṚV. 10.90.3-4) are respectively used to denote the four fractions $\frac{1}{16}$, $\frac{1}{12}$, $\frac{1}{8}$ and $\frac{1}{4}$ respectively.[15]

Regarding the spread of the numerical system from India to other parts of the world, B.B. Dutta observes thus: "From Arabia, the numerals slowly marched towards the West through Egypt and Northern Arabia and they finally entered Europe in the 11th century. The Europeans called them the Arabic notations because they received them from the Arabs. But the Arabs themselves, the eastern as well as the western, have unanimously called them the Hindu figures (Al-Arqan-Al-Hindu)".[16] The word 'cipher' ('sifr' in Arabic) given by the Arabs is the translation of *śūnya* which means empty position.

The discovery of *śūnya* made by the Indian thinkers is the productive point of indifference between positive and negative numbers. As Prof. Halstead observes: "The importance of the creation of the zero mark can never be exaggerated. This giving to airy nothing, not merely a local habitation and a name, a picture, a symbol, but helpful power, is the characteristic of the Hindu race whence it sprang. It is like coining the *Nirvāṇa* into dynamos. No single mathematical creation has been more potent for the general on-go of intelligence and power".[17]

The literal description of zero means a person or thing with no importance or independent existence.[18] The word *śūnya* (*Amarakośa* 3.1.56) in Vedic literature has derived from the word *śūnā (śūnā + yat*). The synonyms of zero *randhra*, *tuccha* (*Amarakośa* 3.1.56), *kṣudra (AV.* 19.23.21), *rikta* (*Amarakośa* 3.1.56) project a conception of nothingness while the synonyms *vyoma (ṚV.* 1.164.35), *diba* (ṚV. 1.22.20), *ākāśa (Amarakośa* 1.2.1), *antarikṣa (Amarakośa* 1.2.1), *jaladhara-patha,* infinite expanse of sky, synonym *pūrṇa (Bṛhadāraṇyakopaniṣad* 5.1.1) means full, and *ananta* (*Amarakośa* 1.2.1) means infinite and synonyms *draviṇam* (*Amarakośa* 3.5.22) and *balam (Amarakośa* 3.5.22) mean strength, vigour and force. In the *Atharvaveda,* one could see the recognition of the negative number in it and its reference to zero. Zero is called *kṣudra* (XIX.22.6) meaning 'trifling'; the negative number indicated by the epithet *anṛca* (XIX.23.22), meaning 'without a hymn'; and the positive number by *ṛca* (XIX.23.1), meaning 'a sacred verse'. These designations were replaced in later times by *ṛṇa* (debt) and *dhana* (asset).[19] Piṅgala (200 B.C.) describes a sūtra *'rūpe śūṇyam'* [or place a zero (*śuñuya*) when unity is subtracted from index or power] among the four *sūtras,* he gives for the computation of 2^n.[20]

The Pāṇinian system has been claimed recently to contribute to the concpt of zero in mathematical sense.[21] It is even said that "He was the first man to use mathematical concept of zero before mathematicians accepted it".[22] The concept of *māyā* and void in the Vedic philosophy, the *śūnyavāda* theory of Buddhists propounded by Nāgasena and Nāgārjuna and the *abhāva* of Nyāya system are all quoted for the genesis of zero. The Advaita Vedānta of Śaṅkarācārya and the Vīraśaiva Śaraṇas call the Highest Principle *śūnya*. But there is a fundamental difference between them in the interpretation of the word. The *śūnya* or *śūnyatā* of the Buddhists is the absolute and ultimate principle, but it is impermanent (*aśāśvata*), whereas that of the Advaita Vedānta and Vīraśaiva Śaraṇas is permanent (*śāśvata*), eternal (*nitya*) and real (*satya*).

The form of zero as a *bindu* or dot is suggested in the Vedic literature. The words *kṣudra* and *randhra* also figure the symbol of zero.[23] The earliest evidence of the symbol for zero is found in the Bakhshālī Manuscript (A.D. 400).[24] A.K. Bag is of the view that "The concept of 'nothingness' or 'void' which is inherent and instinctive to man of all ages found its expression in the philosophical and literary works of all ancient civilised nations. The feeling of necessity to give a symbolical expression of this concept in mathematical calculation which would permit the representation of a number of only high order without any limit led ultimately to the discovery of a symbol for what is now known as zero".[25]

The *śūnya* could be compared to the Brahman which is the minutest of the minute and greatest of the great "*aṇor aṇīyān mahato mahīyān*" (*Kaṭhopaniṣad* 1.2.20). Similarly *śūnya* or zero itself signifies absence when placed independently but it represents fullness when it is placed in the decimal system of numeration (placing of zeros on the right-hand side of a number increases the value to an infinite step).[26] The word *'śūnya'* is identical with the Upaniṣadic word *'pūrṇa'* found in the following *śāntipāṭha*:

pūrṇam adaḥ pūrṇam idaṃ pūrṇāt pūrṇam udacyate/
pūrṇasya pūrṇam ādāya pūrṇam evāvaśiṣyate//

ABBREVIATIONS

AV *Atharvaveda*
SB *Śatapatha Brāhmaṇa*
TB *Taittirīya Brāhmaṇa*
TS *Taittirīya Saṃhitā*
VJ *Vājasaneyī Saṃhitā*
YV *Yajurveda*

REFERENCES

1. R.C. Gupta, 'Who invented the zero?', *Gaṇita Bhāratī*, 17 (1995), p. 58.
2. Ed. by Bangali Baba, Motilal Banarsidass, Delhi, 1976, p. 70.
3. *paṅktiviṃśatitriṃśaccatvāriṃśatpañcāśatṣaṣṭisapatatyaśītinavatiśatam, Aṣṭādhyāyī,* V.1.59.
4. *ekasmai svāhā dvābhyāṃ svāhā tribhyaḥ svāhā caturbhyaḥ svāhā ṣaḍbhyaḥ svāhā saptabhyaḥ svāhā 'ṣṭābhyaḥ svāhā navabhyaḥ svāhā pañcadaśabhyaḥ svāhā ṣoḍaśabhyaḥ svāhā saptadaśabhyaḥ svāhā 'ṣṭādaśabhyaḥ svāhaikān-na-viṃśatyai svāhā navaviṃśatyai svāhaikān-na-catvāriṃśate svāhā navacatvāriṃśate svāhaikān-na-ṣaṣṭyai svāhā navaṣaṣṭyai svāhaikān-nāśītyai svāhā navāśītyai svāhaikān-na-śatāya svāhā śatāya svāhā dvābhyāṃ śatābhyāṃ svāhā sarvasmai svāhā// TS.* VI.2.11//

   Sāyaṇabhāṣya for this portion reads thus : *ekasmai svāheti. atra tv ekaśabdena sarvātmakaḥ prajāparir ucyate. ata eva brāhmaṇe samāmnāsyate — 'ekamasmai svāhety āha. prajāpatir vā ekaḥ. taṃ evā 'pnoti'iti. asya ca sarvātmakatvād ye ye dvitvādisaṃkhyāviśiṣṭāḥ padārthās te sarve 'pi prajāpati-rūpāḥ tasmāt prajāpatirūpān etān sarvān padārthān uddiśyedaṃ svāhutam astv ityevaṃ yathāmnātaṃ viṃśad āhutīr juhūyāt. anukto padārthasaṃgrahārthaṃ sarvaśabdaḥ.*
5. The *Sāyaṇabhāṣya* of this *anuvāka* reads thus : *ekā ca mā iti. ekādiśabdāḥ saṃkhyāparāḥ . . . "yugmaṃ ayuje juhoti mithunatvāya"* (*saṃ. kā.* 5.4.8) *iti. catasraś ca me 'ṣṭau ca ma ityādikaṃ samasaṃkhyārūpaṃ yugmam. ekā ca me tisraś ca me ityādi viṣamasaṃkhyārūpam ayugmam, rāśidvayaṃ strīpuruṣavat parasparavilakṣaṇatvān mithunatvāya saṃpadyate. rāśidvaye 'pyuttarottarādhikyaṃ praśaṃsati — "uttarāvatī bhavato 'bhikrāntyai"* (*saṃ. kā.* 5.4.8) *iti. ekatvād uttaraṃ tritvaṃ, tasmāccottaraṃ pañcatvaṃ, tathā catuṣṭvāduttaram aṣṭattvam. tasmāścottaraṃ dvādaśatvam. evam ubhayoruttarāvattvenābhito vyāptir bhavati. yugmāyugmasaṃkhyāciśabdānāṃ tātparyaṃ darśayati-*

   "ekā ca me tisraś ca ma ity āha devachandasaṃ vā ekā ca tisraś ca manuṣyachandasaṃ catasraś cāṣṭau ca devachandasaṃ caiva manuṣyachandasaṃ cāvarundhe" (saṃ. kā. 5.4.8)

   viṣamasaṃkhyayā devānām priyāṇi cchandāṃsi niṣpādyante, samasaṃkhyayā tu

manuṣyapriyāṇi. atas tadubhayaṃ prāpnoti. rāśidvaye 'py uttarottarasaṃkhyāvṛddher evāvadhiṃ praśaṃsati-
"ā trayastriṃśato juhoti trayastriṃśad vai devatā devatā evāva rundha ā'ṣṭācatvāriṃśato juhoty aṣṭācatvāriṃśadakṣarā jagatī jāgatāḥ paśavo jagatyaivāsmai paśūnava rundhe" (*saṃ. kā.* 5.4.8) *iti.*

6. *Śivasaṅkalpopaniṣat*, 13. published in the *'Un-Published Upanishads'*, ed. by the Pandits of Adyar Library, Adyar Library Series No. 14, Madras, 1933, p. 320.
7. Also see *Vājasaneyīsaṃhitā*, XVII, 2; *TS.* IV.40.11.4; VII.2.20.1.
8. *śatāya svāhā sahasrāya svāhā 'yutāya svāhā niyutāya svāhā prayutāya svāhā 'rbudāya svāhā nyarbudāya svāhā samudrāya svāhā madhyāya svāhā 'ntāya svāhā parārdhāya svāhoṣase svāhā vyuṣṭyai svāhodeṣyate svāhodyate svāhoditāya svāhā suvargāya svāhā lokāya svāhā sarvasmai svāhā. //*VI.2.20//
9. Gita Press, Gorakhpur, 1992, p. 1465.
10. Ed. by S. Lefmann, Halle, 1902, p. 146.
11. B. Datta, *The Cultural Heritage of India*, Vol. VI, *Science and Technology*, The Ramakrishna Mission, Calcutta, p. 35.
12. The seven steps at a marriage. The bride and bridegroom walk together seven steps, after which the marriage becomes irrevocable.
13. See also J. Gonda, 'Triads in Vedic Ritual', *Ohio Journal of Research Studies*, 2.2 (Ohio 1974), pp. 5-24.
14. Cf. *dakṣiṇā gāyatrī-sampannā brāhmaṇasya (Kātyāyana-śrautasūtra* 22.11.21): Com. *Gāyatryakṣarasamānasaṅkhyāḥ caturviṃśatir gāvo dakṣiṇā.*
Jagatyā rājñaḥ (Kātyāyana-śrautasūtra 22.11.22). Com. *Jagatyakṣarasamānasaṅkhyā aṣṭacatvāriṃśad gāvo bhavati.*
See also K. Menninger, *Number Words and Number Symbols*, MIT Press, Cambridge, Mass., 1970.
15. see also K. Menninger, *Number Words and Number Symbols*, MIT Press, Cambridge, Mass., 1970.
16. 'Indian Invention of Decimal System and Number Zero' by Mukesh Dutt Sharma, in *Vedic Path*, 44.1 (June 1982), p. 35.
17. *On the foundation and technique of Arithmetic*, Chicago, 1912, p. 20.
18. For example see this ancient saying *'avidyaṃ jīvanaṃ śūnyam'*: 'Without learning the life is void (or zero)'.
19. R.N. Mukherjee, 'Background to the discovery of the symbol for zero', *Indian Journal of History of Science*, 12 (1977), pp. 225-31.
20. R.C. Gupta, 'Who invented the Zero?', *Gaṇita Bhāratī*, Vol. 17, p. 57.
21. M.D. Pandit, *Zero in Pāṇini*, Poona Uni., Pune, 1990.
22. *Ibid.*, p. 23.
23. R.N. Mukherjee, *op. cit.*, p. 225.

24. Vide Hoernle, A.F.R., 'The Bakhshālī Ms.', *Indian Antiquary*, 17, pp. 33-48, 1888; Kaye, G. R., *The Bakhshālī Ms.*, Parts I and II, New Imperial Series No. 43, Calcutta, 1927; Dutta, B. 'The Bakhshālī Mathematics', *Bulletin of the Calcutta Mathematical Society*, 21 (1929), pp. 1-60.
25. A.K. Bag, *Mathematics in Ancient and Medieval India*, Delhi, 1979, p. 67.
26. R.N. Mukherjee, 'Background to the Discovery of the Symbol for Zero', p. 227.

Word and Alphabetic Numerical Systems in India

K.V. SARMA

NEED FOR NUMERICAL SYSTEMS

The Setting

Word or object numerals (*bhūta-saṅkhyā*) and Alphabet or Letter numerals (*akṣara-saṅkhyā*) formed a natural development aimed at the concise expression of numbers with ease in the fast-growing scenario of scientific development in ancient India, in the fields of astronomy and mathematics in particular. Naturally enough, in the actual formulation and expression of these two numerals were involved certain characteristics of the language of Sanskrit which functioned as the primary vehicle of culture in India from the earliest times. Metric form and brevity of expression constituted two of these characteristics. In fact, the extraordinarily large quantum of literature produced in early India had been couched in the metric or aphoristic form which conduced to their study by rote and preservation in memory. Then again, abridgements and abbreviated terms were quite common in the range of Sanskrit literature on account of the natural propensity and adaptability of that language. An example or two might be cited in illustration. In the *Vedāṅga-Jyotiṣa* the twenty-seven asterisms of Indian astronomy are denoted by abbreviation (*ṚV-VJ*. 14-15; *YV-VJ*. 17-18).[1] Thus *jau* stands for the asterism *Āśvayujau* and *drā* for *Ārdrā*; and so on. In Sanskrit grammar the abbreviation *ac* represents all the thirteen vowels and *hal* represents all the consonants. Herein lies the genesis of the Word and Alphabetic numerals which are profusely employed in Indian astronomy and mathematics and, to a lesser extent, in other disciplines as well.

The Numerical Systems

Several numerical systems based on full-fledged words and on individual letters of the alphabet had been devised in India at different periods, in different places to cater to general and specific needs. Word-numerals were evolved earlier and are traceable to Vedic literature. They occur in several modes such as single words, groups with place value and in the reverse where numbers denote objects. Short numerals like *dvi* and *tri* are found to intermingle with word numerals. Alphabetic numerals, which came later, have been developed by assigning different values to vowels, consonants, conjunct consonants and their components and also the placement thereof, mostly with some sequence and logic behind them. In some of the systems, however, the assignment of values appears to be purely arbitrary and *ad hoc.*

WORD-NUMERALS

Nature of Word-numerals

The word-numeral, or, to be more exact, 'object-numeral' (*Bhūta-saṅkhyā*) refers to a system of numerical notation where a word denoting an object, concept, idea or group implies, besides its normal meaning, a number also by dint of meaning, convention or usage and, for that reason, that word is used to denote the number in question. Thus the number 4 can be denoted by one or the other of the words *veda, varṇa, āśrama* or *yuga* which are all groups of four. The number 2 can be denoted by any word which exists only in pairs like the eyes, arms and the like. *Rudra* denotes 11 since the *rudra*-deities are eleven in number. In the same way, *rada* (teeth) would denote 32 and *nakṣatra* (asterisms) 27. The synonyms of a word-numeral can also be used to indicate the number in question. The phenomenal number of synonyms, each of varying lengths and containing differing number of syllables with differing alternations of short and long syllables, made these word-numerals extremely amenable to be fitted in appropriately into verses in different metres.

In actual usage these word-numerals as also ordinary numerals, which are also often found interspersed, are treated just as any other word with the addition of all types of grammatical complements like gender, affixes, suffixes and the like so that they merge into the verse and do not appear to be extraneous interpolations. The words in the verse other than the word-numerals serve to explain the procedures and computations or other matters involved, as the case may be. Then again, a word-numeral standing by itself denotes only itself, without place value for instance, *atyaṣṭi*, being the name of the metre having 17 syllables in a quarter, would denote the number 17. Long numbers are formed by appropriate word-numerals being placed one after the other, with place value, and read from right to left according to the principle *aṅkānām vāmato gatiḥ,* that is, 'the way of placing the numerals is from right to left'. Thus the expression *kha-rāma-guṇa-candraiḥ guṇayet* would mean 'Multiply by 1330'. The words which are conventionally employed as word-numerals number more than 800 well beyond their enumeration here. They have been compiled and presented as an Appendix by K.S. Shukla to his edition of *Vaṭeśvarasiddhānta*, Part I (INSA, New Delhi, 1986). (Extracted below as Appendix I).

It is necessary to note that it is only in India and in the Sanskrit language among all the world languages that word-numerals have been profusely used to represent numerals, from as early as before the Christian era.

Word-Numerals in the Reverse

It is interesting that sometimes certain objects which occur only in numbered groups are denoted by the total number of items in the respective groups. For instance *dvādaśa* (12) is used to denote the 'year' which has twelve months and *sapta* (7) to indicate objects which occur in groups of seven like the oceans and planets.[2]

PROVENANCE AND POPULARITY

The Vedic Texts

The use of word-numerals has been an all-India phenomenon from very early times, especially in North India. On account of Sanskrit having been the main language in which literature was produced in ancient India, writers in several disciplines have adopted the device of the word-numeral notation to depict numbers in their works. However it was authors of texts on astronomy and mathematics that benefited most since those two disciplines were centred on numbers.

From among the Vedic texts where word-numerals occur might be mentioned *Kātyāyana-Śrautasūtra* and *Lāṭyāyana-Śrautasūtra* where the words *gāyatrī* and *jagatī*, being metres having specific numbers of syllables, are used to denote 24 and 48, respectively.[3]

The four fractions 1/16, 1/12, 1/8 and 1/4 are denoted in Vedic literature respectively by the words *kalā (ṚV.* 8.47.17), *kuṣṭha (YV. Maitrāyaṇīya Saṃhitā* 8.7.7), *śapha (ṚV.* 8.47.17) and *pāda (ṚV.* 10.90.3-4). The word-numeral *rūpa* (1) occurs in the *ṚV-Vedāṅga Jyotiṣa* 31.

In the *Agnipurāṇa* word-numerals occur prominently in several of the chapters like *Kālajñāna* (ch. 122).[4]

General Disciplines

Several of the texts in *Tantraśāstra*, for instance, *Sāradātilaka*,[5] *Tantrasamuccaya* of Nārāyaṇa[6] and *Mayamata*[7] make use of the word-numeral notation. In music the *Caturdaṇḍiprakāśikā* of Veṅkaṭamakhin designates the twelve divisions of his musical scheme as *indu* (1), *netra* (2), *agni* (3), *veda* (4), *bāṇa* (5), *ṛtu* (6), *ṛṣi* (7), *vasu* (8), *brahmā* (9), *dik* (10), *rudra* (11) and *āditya* (12). Coming to architecture, the *Śilparatna* uses the word-numerals in measurements relating to the construction of the eight-storey edifice.[8]

Inscriptions

Inscriptions too have their dates inscribed in word-numerals. The earliest instances are to be found in two Sanskrit inscriptions dated *śaka* 526 (A.D. 604) and 547 (A.D. 625) discovered in Java at which period it was an Indian colony.[9] Inscriptions and copperplates too have their dates inscribed in word-numerals.[10]

Colophons of Manuscripts

Colophons of texts and post-colophonic statements by scribes of manuscripts too give their dates, at times, in word-numerals.[11]

Astronomy and Mathematics

The greatest impact of the system was on the twin disciplines of mathematics and astronomy whose mainhold was numbers. These two sciences underwent an efflorescence in India from about the beginning of the Christian era from when on numerous works came to be produced. Practically all the major astronomers and mathematicians of the period took full advantage of the system for depicting, in simple graceful verses, the fundamental parameters and computations involved in their works.[12] These included, among others, the promulgators of the five *siddhāntas Paitāmaha, Vasiṣṭha, Pauliśa, Romaka* and *Saura* epitomised in the *Pañcasiddhāntikā* of Varāhamihira, Bhāskara I, Brahmagupta, Śrīdhara, Bhaṭṭotpala, Śrīpati, Vaṭeśvara, Bhāskara II and a host of others, mainly from North India, who came later. Some of the prominent scholars of South India like Parameśvara of Vaṭaśreṇi and Nīlakaṇṭha Somayāji also used the system extensively.[13]

ALPHABETIC NUMERICAL SYSTEM

Kaṭapayādi-I System

Nature of the System

Far more facile and much more versatile than the word-numeral system is the Kaṭapayādi-I alphabet system developed in Kerala. Here individual letters of the alphabet are adapted in a logical manner to represent the digits 1 to 9 and zero, and arranged in the traditional way of right to left to form numbers. The virtuosity of the system stems from several factors. First, each of the digits and zero could be represented by any of several specified letters. Conjunct letters and single letters could have any vowel attached to them and can take any case-ending whatsoever and function accordingly without any alteration to their values. These two aspects enabled one to weave words and expressions which could retain their normal meanings, form meaningful phrases and merge into metrical verses and, at the same time, represent the intended numbers as well. To cite an example, the word *āyurārogyasaukhyam*, with which the well-known work *Nārāyaṇīyam* of the devotional poet of Kerala Nārāyaṇa Bhaṭṭa closes, sets out the date of the completion of the work in Kali days as 17,12,211 by the Kaṭapayādi notation. Besides this the expression gives, along with the other words of the verse, its normal meaning, being the prayer of the author for longevity (*āyus*), health (*ārogyam*) and happiness (*saukhyam*). The significant advantage of the system lies in the fact that the short and crisp numerals formed carry contextual meanings also through full-fledged and grammatically correct expressions which make them easy to memorise and render them not liable to corruption in speech or in writing. Also, if-ever a doubt arose the same could be checked through meaning and grammar.[14]

The System Explained

Conversion of even long numbers into Kaṭapayādi and the formulation of tables of different types can be effected with ease. The art is practised even today by the astronomers and almanac makers of Kerala in their works.[15] The several almanacs issued every year in Malayalam by Kerla astronomers and the horoscopes for new-born babies prepared by the astrologers of the land have dates set out in the Kaṭapayādi system. The reason for this lies in the simplicity and straightforwardness of the system. Here all vowels are taken as zero and those affixed to consonants are ignored. The consonants alone are taken to represent the digits and are taken in the order in which they occur in the Sanskrit alphabet. In the case of conjunct consonants, the last component alone is taken as a digit. Thus the ten letters from *ka* to *ña* represent 1 to 9 and zero. Then the ten letters *ṭa* to *na* are again taken to represent 1 to 9 and zero, and the next five letters *pa* to *ma* are taken as 1 to 5 and the last nine letters *ya* to *ḷa* are taken as 1 to 9. It is to be noted that *ḷa* is a distinct letter of the alphabet in the Malyalam and other South Indian languages as in the Veda. The rule for Kaṭapayādi-I is given by the verse :

> *na-ñāv acaś ca śūnyāni, saṅkhyāḥ kaṭapayādayaḥ/*
> *miśre tūpāntya-hal saṅkhyā, na ca cintyo hal asvaraḥ//*
> (*Sadratnamālā* of Śaṅkaravarman, 3.3)

'*na, ña* and the vowels are zero. The letters (of the consonant groups) commencing with *ka, ṭa, pa* and *ya* are digits. In conjunct letters the last consonant is to be taken as the digit. A consonant not attached to a vowel is to be ignored.'

Date of the System

Āryabhaṭa (born A.D. 476), who designed his own system of alphabet numerals, was conversant with the Kaṭapayādi system as shown by his commentator Sūryadeva Yajvan while explaining the word *kāt* in *Āryabhaṭīya* 1.2 (see edn. INSA, New Delhi, 1976, p. 10). Vararuci, the

Kerala astronomer who prepared the table of the moon's daily angular motion (*candravākyas*) using the Kaṭapayādi system lived in the fourth century A.D. Another early work which makes use of the Kaṭapayādi notation is the *Upadeśasūtra* or *Jaiminisūtra* on astrology by Jaimini.[16] However it is difficult to accept the identification of this author with Jaimini, author of the *Mīmāṃsāsūtra*s and the statement that "the system probably existed even as early as the fifth century B.C." and even earlier since "The *Mahābhārata* is called *Jaya* 'because' *jaya* in Kaṭapayādi system means eighteen".[17]

Provenance and Popularity

The popularity of the Kaṭapayādi system in Kerala has been phenomenal. This is not surprising because one of the texts prescribed for children when they commenced their studies according to the traditional method, a system which extended from very early times and continued till recent times, was the *Candravākyas* (Table of Moon-sentences) composed in Kaṭapayādi by Vararuci. The acquaintance with the Kaṭapayādi notation, imbibed when young, enabled writers of every discipline to use this facile system of expressing numbers whenever there was a need. Thus whatever be the subject of a text, *Veda, Philosophy, Dharmaśāstra, Tantra, Śilpa, Vaidya, Vyākaraṇa, Saṅgīta* or *Kāvya*, the authors often indicated the Kali dates of completing their works in Kaṭapayādi and much more so the scribes of the manuscripts in their post-colophonic statements. Vedic students used to identify, remember and recollect the chapter and verse of the texts through Kaṭapayādi abbreviations. Nārāyaṇa Bhaṭṭa of Melputtūr has documented the *aṣṭaka* divisions and *varga* sections of the *Ṛgveda* through the Kaṭapayādi notation in a short work called *Sūktaślokāḥ*.[18] It needs to be added that in its non-technical normal sense the entire work forms a eulogy on the Mother Goddess. Wordy acrobatics using the system were also not absent among Kerala poets.[19] The system has been adopted by the discipline of music as well. The *Caturdaṇḍiprakāśikā* of the 17th century musicologist Veṅkaṭamakhin who has classified the *rāgas* of South Indian Carnatic music gives the names of the 72 *rāga*s in such a manner that the first two letters of each name gives the number of the *rāga* in the Kaṭapayādi system. The rest of the letters in the names give some characteristic of the relevant *rāga*. For example, No. 9 is termed *'dhuni'-bhinnaṣaḍja*; No. 28 is *'hari'-kedāragaula*.[20]

However, it was in the disciplines of astronomy and mathematics, which revolved around numbers, that the fullest use of the Kaṭapayādi numerals was made, especially in the formation of the constants, parameters, multipliers, divisors and the large number of sine tables involved in different computations. It is instructive to note here that in Kerala periodical revision of the parameters and tables had been directed by competent astronomical authorities like Parameśvara of Vaṭaśreṇi (A.D. 1360-1455),[21] Nīlakaṇṭha Somayājī (1444-1545)[22] and Jyeṣṭhadeva (1500-1610).[23]

It has to be noted that under the influence of such guiding instructions a large number of astronomical manuals (*Karaṇa*-texts) and tracts of different epochal dates, with differing parameters and sine tables (*jyā*s) have come to be composed in the land. Apart from these a large number of texts catalogued merely as '*jyā*s' are preserved on the shelves of manuscript repositories in Kerala.[24] This manuscript wealth, wherein all numbers are represented in the Kaṭapayādi system, would be a pointer to the wide prevalence of the system in Kerala.

From Kerala the Kaṭapayādi system has flowed over to the neighbouring region of Tamilnadu and to a lesser extent to the Telugu and Kannada-speaking regions. Almanacs in Tamilnadu are generally called *Vākyapañcāṅga*, 'Almanac based on mnemonic phrases'.

In their computation, while true moon is calculated using the 248 Moon-sentences of Vararuci, the true planets are computed using 2075 planet-phrases which go by the name *Samudravākya*s or *Maṇḍala-vākya*s, all set out in Kaṭapayādi.[25] Still another astronomical *Karaṇa* text produced in Tamilnadu is the *Tantraratna* of Gopendra Tippabhūpāla (fl. A.D. 1475) who has composed his own tables of 248 *Candravākya*s correct to the seconds in Kaṭapayādi notation.[26]

The Jain School of mathematics of the Karnataka region involves substantial mathematical material even in their religious and philosophical works like the *Trilokasāra, Gommaṭasāra* and *Labdhisāra* of Nemicandra Siddhāntācārya (9-10 cent.) wherein full use of the Kaṭapayādi-I system is made; vide SN. on 'Symbolism and List of working symbols', on pp. 59-61 of vol. I of *Labdhisāra,* ed. by L.C. Jain (Katni, 1994).

ALPHABET-NUMERALS OF ĀRYABHAṬA I

Nature of the System

Āryabhaṭa (b. A.D. 476), astronomer and mathematician who flourished at *Kusumapura* (Patna), was aware of the Kaṭapayādi-I system[27] wherein full-fledged letters, single and conjunct, with vowels attached, represented a single digit. Obviously, with a view to prune that system further, Āryabhaṭa devised another system in which a consonant-pure, i.e. a consonant without a vowel attached, represented not only single digits but also double-digited numbers. A conjunct consonant was to be the sum of all the numbers of its components. Each vowel indicated an odd and an even place, in the ascending order, as to where a digit was to be placed to form a number.

The System

Āryabhaṭa enunciates the system in the verse:

> *vargākṣarāṇi varge'vargākṣarāṇi kāt, ṅmau yaḥ/*
> *khadvinavake svarā, varge 'varge navāntyavarge vā//*
>
> (*Āryabhaṭīya* 1.2)

'While deciphering numbers from the alphabet-numerals of this system the *varga* letters (i.e. the 25 letters *k* to *m* being five each of the five *varga*s, *k, c, ṭ, t* and *p* of the Sanskrit alphabet, with the assigned values 1 to 25) should be written in the *varga* places (or odd places, i.e. 1st, 100th, 1000th, etc. which are *varga*s or squares) and the (eight) *avarga* letters (*y* to *h*) with the assigned values 30 to 100 should be written in the *avarga* (or even) places; *ṅ* and *m* (i.e., 5 and 25) together make *y* (or 30). Assign eighteen zeros (to represent 18 notational places, two each to the nine) vowels (*a* to *au*) which will indicate the *varga* or *avarga* places (where the letters to which they are attached should be placed). For numbers bigger than 18 notational places repeat with another set of vowels or with some other difference.'[28]

Continuing, Āryabhaṭa gives in his numeral system the eastward revolutions of the planets:

> *yugaravibhagaṇāḥ khyughṛ, śaśi*
> *cayagiyiṅuśuchlṛ, ku ṅiśibuṇlṛṣkhṛ prāk/*

śani ḍhuṅvighva, guru khri-
cyubha, kuja bhadlijhnukhṛ, bhṛgubudhasaurāḥ//
(*Āryabhaṭīya* 1.3)

Thus the eastward revolutions per aeon of the Sun are *khyughṛ* or 43,20,000; of the Moon *cayagiyiṅuśuchlṛ*, 5.77,53,336; of the Earth *ṅiśibuṇlṛṣkhṛ* or 1,58,22,37,500, of Saturn *ḍhuṅvighva* or 1,46,564, of Jupiter *khricyubha* or 3,64,224; of Mars *bhadlijhnukhṛ* or 22,96,824; of Mercury and Venus, the same as that of the Sun.

Though the chronograms formed by this system are formed logically and are extraordinarily brief, this advantage is offset completely by their not being amenable to be freely pronounced, copied with confidence, memorised with ease, not suffixed with a case-ending nor fitted into a verse or formula. Moreover they lack elegance and have no literary sense whatsoever and so it is practically impossible to be checked or corrected in case of doubt. It would seem that the system had no takers. Neither has any later astronomer employed this numeration nor is it found in any inscription.

KAṬAPAYĀDI-II SYSTEM OF ĀRYABHAṬA II

The System Explained

Āryabhaṭa II who wrote his *Mahāsiddhānta*[29] in about A.D. 950 modified the Kaṭapayādi-I system with some innovations and devised the Kaṭapayādi-II system for use in his work. The rule is :

rūpāt kaṭapayapūrvā varṇā varṇakrameṇa bhavanty aṅkāḥ/
ñanau śūnyam, prathamārthe ā chede, ai tṛtīyārthe//
(*Mahāsiddhānta* 1.2)

'(The four groups of) consonants beginning with *k, ṭ, p, y,* each from 'one' (*rūpa*), function as digits; *ñ* and *n* represent zero. (Chronograms expressed) in the nominative case (plural) and in the instrumental (plural) drop their (ending *visarga*s) and read with an ending *ā* and *ai*, respectively.'

An examination of the presentation of numbers in the *Mahāsiddhānta* provides some more information about the system. Thus, while accepting the values of the consonants, taking *ñ* and *n* as zero and the vowels attached to consonants as of no significance as in Kaṭapayādi-I, the new system gives values to all the components in a conjunct consonant. Thus *kya* means 11, not 1 as in Kaṭapayādi-I.

The deletion of the *visarga* is obviously intended to obviate the possibility of the *visarga* turning into the letters *r, ś, ṣ* and *s* when euphonic combination (*sandhi*) takes place between the *visarga* and an amenable consonant following it (in an adjacent chronogram), thereby changing the number intended by the chronogram. For example, the chronogram *dāsāḥ cātu*, will remain as such if the words are given as *dāsā cāṭu* but will become *dāsāścāṭu* if *sandhi* is allowed. Then the digit for *ś* will get included in the choronogram and render the number different. It might be noted that in Kaṭapayādi-I such a contingency does not arise since chronograms are all composed by means of single meaningful words. It is also to be noted that the above modification is prescribed only for nominative and instrumental plurals for the reason that chronograms are given only for setting out numbers in the nominative and to instruct their multiplication and division in the instrumental.

The most important modification, however, in this system lies in that the chronograms are given with the digits set out from left to right as sometimes practised in early days and always so in modern times. Thus *manana* is 500 and not 005 which it would be according to Kaṭapayādi-I.

Limitations of the System

Obviously enough, Āryabhaṭa II enunciated his system of Kaṭapayādi with a view to simplifying it and incorporated an idea or two from Āryabhaṭa I's system. However, neither of these devices has been of much help on account of the inanity of the chronograms that have been concocted by the author.[30]

If the chronograms set in the numerals of Āryabhaṭa-I are not pronounceable and teeth-breaking on account of heavily packed conjunct consonants, those of Āryabhaṭa II are replete with liquid sounds and consecutive repetition of the same letters. See, for example, the revolution of the sun given by the chronogram *ghaḍaphenenanunīnāḥ* (1.7) and the number of civil days by the chronogram: *yaṇisasudhīkosīmeḍhīkhunīnonāḥ* (1.13). Here, there is always the possibility of tripping a letter or two not only in pronunciation but also in writing. In both systems, it is easy to err and difficult to check, with the result that both these letter-numerals had practically no takers.

PALI KAṬAPAYĀDI SYSTEM

In Pali texts the Kaṭapayādi-I numeration is employed with a slight difference. The Pali alphabet does not have the two letters *ś* and *ṣ* but has an additional letter *l'*. Hence the Kaṭapayādi numerals used in Pali are identical with Kaṭapayādi-I except that after $v = 4$, it is $s = 5$, $h = 6$ and $l' = 7$.

KAṬAPAYĀDI IN THE 64 ARTS

From among the well-known sixty-four Arts, the 46th *Mlechhita-vikalpa* ('Manipulation with incoherent speech') includes several manipulations. Of these *Gūḍhalekhya* ('Cipher writing') uses Kaṭapayādi in a selective manner with facilities for coding and decoding, employing the formula:

vedāṅgasārakṣapaṇāḥ sthāṇur lokahito varaḥ/

Coding is done using the first part of the formula by rewriting the letters of the passage which requires to be coded in the order 4 (*ve*), 8 (*dā*), 3 (*ṅga*), 7 (*sā*), 2 (*ra*), 6 (*kṣa*), 1 (*pa*) and 5 (*ṇāḥ*). Thus a secret message which reads: *rakṣakaḥ sukham svapiti* ('The guard is sleeping soundly') will be coded as: *sutikaḥ pikṣasva rakham*, which is a nonsensical jumble of letters. When this is decoded by rearranging the letters in accordance with the second part of the formula, 7 (*sthā*), 5 (*ṇur*), 3 (*lo*), 1 (*ka*), 8 (*hi*), 6 (*to*), 4 (*va*), 2 (*raḥ*), the original passage is obtained.

Since the formula can accommodate passages only up to eight-letter codes, for a passage of more letters, the coding and decoding shall have to be repeated for every eight letters.[31]

FOLIO NUMBERING IN MANUSCRIPTS

Nature of the System

Folios of manuscripts written on paper, *bhūrja-patra* and similar material are generally numbered by means of the normal decimal digits. The marking is made on the reverse side of the folios on the upper right hand corners, generally in line with the first row of the written matter. In the case of palm leaf manuscripts of South India, however, the folio numbers are inscribed with letter-symbols in the script in which the manuscript is written, on the middle of the left hand margin on the front side of the folios. This system of folio-numbering is known as *akṣara-pallī* (letter system) as contrasted with *aṅka-pallī* (digit system) wherein markings are done with digits.

Table of Letters used

On the basis of a detailed examination of a large number of manuscripts inscribed in different scripts and belonging to different periods, deposited in several institutions, Pt. Gaurishankar H. Ojha has prepared a fairly exhaustive table of letter-forms used for indicating folio-numbers in manuscripts.[32] Ojha's table adapted below gives the list of his findings with some later additions included:

1. *e, sva, oṃ, śrī*
2. *dvi, sti, na, nna*
3. *tri, śrī, maḥ*
4. *ṅka, rṅka, ṅkā, eka, erka, rṣka, ṣkra, rph*
5. *tṛ, rtṛ, rtṛā, hṛ, ṇr, jhra*
6. *phra, rphra, ṛphra, ghra, bhra, rpu, vyā, phla, hā*
7. *gra, grā, rgbhrā, rggā, bhra, rgrā*
8. *hra, rhra, rhrā, dra, pra*
9. *oṃ, rūṃ, ru, uṃ, ūṃ, a, rnum, dre*
10. *lṛ, ḷa, ṇṭa, ḍā, a, rptā*
20. *tha, thā, rtha, rdhā, gha, rgha, pva, va*
30. *la, lā, rla, rlā*
40. *pta, rpta, ptā, pna*
50. *ba, e, i, ṇū*
60. *cu, vu, ghu, thu, rthu, thū, rthū, rgha, rghu, ghu, tru*
70. *cu, cū, thū, rthu, rghū, rmta, tru*
80. *pu, cha*
90. *ṇa*
100. *su, sū, rsṛ, ā, lū, rghū, ña*
200. *su, sū, rsu, ā, lū, rgh*
300. *stā, sūa, ñūa, sā, su, suṃ, sū*
400. *sūo, sto, stā.*

It is to be noted that some of these letter-forms accord with normal letters in certain scripts but many of them are odd forms which are not used in any language. Possibly they have been concocted with variations in order to distinguish them from the letters used in the languages. Some of them repeat in different languages with different values. However the question remains as to how these letter-forms which have no apparent sequence or meaning have been selected to represent the sequential numerals. It is clear, however, that *e, dvi* and *tri* which stand for 1, 2 and 3 are the letters commencing the respective numbers, and *sva* = 1, *sti* = 2, and *śrī* = 3, are the letters of the auspicious expression *svasti śrīḥ*. *Śrī* = 1 is also such a word. In all cases, the second, and further letters representing the ten's and hundred's places are inscribed alongside the letter of the unit's place if there be space enough in the margin and otherwise one below the other.

Na-nna-nya-ṣkra scheme of Mss. in Malayalam Script

An alphabetical-numeration scheme systematically followed in numbering folios of the palm leaf manuscripts in Kerala from early days to recent times is the *na-nna-nya-ṣkra* scheme where *na* = 1, *nna* = 2, *nya* = 3, *ṣkra* = 4, *jhra* = 5, *hā* = 6, *gra* = 7, *pra* = 8, *dre* = 9, *ma* = 10, *tha* = 20, *la* = 30, *pta* = 40, *ba* = 50, *tra* = 60, *tru* = 70, *ca* = 80, *ṇa* = 90, *ña* = 100. Often the first folio is numbered with the letter *śrī* and the regular numbering commences from the second folio with *na*.

Folio-numbering in Mss. in Tamil Script

The Tamil language has not developed numerical symbols but uses some letters of its alphabet instead. The same alphabetic numerals are used in manuscripts also. Herein, 1 = *ka*, 2 = *u*, 3 = *na* with a dash at the foot, 4 = *ca* with an upward dash, 5 = *ru*, 6 = *cā* slightly modified, 7 = *e*, 8 = *a*, 9 = *ka* modified, 10 = *ya* rounded, 20 = *u* prefixed to the symbol for 10, 100 = *ḷa* without the initial loop, and 1000 = *ta*. Here again, it is hard to know the reason behind this apparently ad hoc selection of the letters to represent the different numerals.

Folio Numbers in Mss. in Grantha Script

The Grantha script which is used to inscribe Sanskrit texts in the Tamil region, too, has not developed numerical symbols and uses some of the letters of its alphabet to represent the numerals. The same symbols are used in numbering manuscripts also. It follows Tamil numbering with slight variations. Thus, 1 = *ka*, 2 = *u*, 3 = *ṅa* modified, 4 = *ca* modified, 5 = *ru*, 6 = *cā* modified, 7 = *e*, 8 = *a*, 9 = *ka* modified, 10 = *dha*. It might be noted that the other Dravidian scripts have developed their own numeral symbols which are used also in numbering folios in manuscripts.

SEQUENTIAL DEVEPLOPMENT OF THE ALPHABET FOR NUMERATION

Nature of the Scheme

The order of the letters of the alphabet in a language, which must be well known to the users of that language, offered another logical method for alphabetic numeration. The devices of both Āryabhaṭas and the Kaṭapayādi systems utilised this method in part, to the extent they required it. The principle involved is to make a letter represent the serial number of the place occupied by it in the respective alphabets. Several systems with minor variations, have come up following this principle.

Pāṇini's use of Alphabet-Numerals

From among the nearly 4,000 grammatical rules (*sūtra*s) framed by Pāṇini (*c.* 600 B.C.) for the Sanskrit language there are a number of 'head-rules' (*adhikāra-sūtra*s) which govern and overflow to several secondary rules which follow. While explaining *sūtra* 1.3.11, *svaritenādhikāraḥ*, Patañjali in his commentary *Mahābhāṣya* states that Pāṇini used to indicate by a sign the number of secondary rules which would be governed by it. Patañjali's commentator, Kaiyaṭa, explains how this indication is made when he comments on *sūtra* 5.1.30, *dvitripūrvānniṣkāt.* He says that through the indicatory syllable *i* here, which means two, Pāṇini instructs that the next two *sūtra*s would be governed by this *sūtra*. And, it follows that according to Pāṇini *a* = 1, *i* = 2, *u* = 3, and so on up to *ha*, the order of the letters being that given by him in his *Pratyāhāra-sūtra*s, called also *Śivasūtra*s.[33]

Ka-kha-ga-Method

In Kerala some manuscripts number the folios with the consonants of the alphabet and not by the *na-nna-nya-ṣkra* method. Some manuscripts in the Telugu region also do so. In this scheme numbers 1 to 34 are marked by *ka* to *ḷa*, 35 to 68 by *kā* to *ḷā* (*ā* being added to the different consonants in place of (*a*), number 69 to 102 by *ki* to *ḷi* and so on.[34]

Several variations to the above are met with elsewhere. In Burma (now Myanmar) Buddhist manuscripts number the folios as *ka, kā, ki, kī* to *kaḥ* being numbers 1 to 12, *kha, khā* to *khaḥ* (13-24), *ga, gā* to *gaḥ* (25-36) and so on. It is to be noted that in this system, only twelves are used leaving the short and long of *ṛ* and of *ḷ*, which are dropped. In the manuscripts of Sri Lanka (Celyon), however, the folio numbers 1-16 would be represented by *ka* to *kaḥ*, 17-32 by *kha* to *khaḥ* etc. In the case of very long manuscripts when all the consonants have been exhausted, the numbering would restart with 2*ka*, 2*kā* etc.[35]

The methods of numbering folios of manuscripts described above pertain only to South India and the lands where South Indian culture had spread. It is also to be noted that these numerical symbols stand by themselves and are not incorporated into literature like the *Bhūtasaṅkhyā* and Kaṭapayādi systems.

In fine it has to be stressed that the ingenious device of word and letter numerals has facilitated the graceful and memorisable presentation of numbers in Sanskrit in the fields of mathematics and astronomy as much as the actual digits in calculation and computation.

REFERENCES

1. *jau drā ghaḥ khe śve 'hī ro ṣā cin mū ṣa ṇyaḥ sū mā dhā ṇaḥ/*
   re mū ghāḥ svā 'po 'jaḥ kṛ ṣyo ha jye ṣṭhā ity ṛkṣā liṅgaiḥ//
2. *Ṛgveda (ṚV)* 7.103.9 : *devahitam jugupuḥ 'dvādaśasya'*: (They keep the god-appointed order of the twelve-monthed sun).
   Atharvaveda (AV) 1.1.1: *'ye tri-ṣaptāḥ pariyanti viśvam'* (The three 'seven-groups' which go round the earth).
3. Cf. *dakṣiṇā gāyatrī-sampannā brāhmaṇasya* (Kāt. Śs. 22.11.21): Com. *Gāyatryakṣarasamānasaṅkhyāḥ caturviṃśatir gāvo dakṣiṇā.*
   Jagatyā rājñaḥ (Kāt. Śs. 22.11.22). Com. *Jagatyakṣarasamānasaṅkhyā aṣṭacatvāriṃśat gāvo bhavati.* See Lāt. Śs. 9.4.31 for a similar passage and similar comments.
4. See *Agnipurāṇa*, ch. 122 (*Kālajñāna*):

   kālaḥ samāgaṇo, vakṣye gaṇitam kālalabdhaye/
   kālaḥ samāgaṇo 'rka-ghno māsaiś caitrādibhir yutaḥ//1//

dvighno dvighno sa-vedaḥ syāt pañcāṅgāṣṭayuto guṇaḥ/
triṣṭho madhyo vasuguṇaḥ punar vedaguṇaiś ca yaḥ//2//
aṣṭarandhrāgnihīnaḥ syāt adhaḥ saika-rasāṣṭakaiḥ/
madhyo hīnaḥ ṣaṣṭihṛto labdhyauktas tathopari//3//

Here the words *arka* (12), *veda* (4), *vasu* (8), *randhra* (9), *rasa* (6) and *agni* (3) are all numerals.

5. Cf. *rudra* (12) *saṅkhyākṣaro manuḥ*. (7.96)
6. Cf. *tattvā-ṣṭi-tattva* (25, 16, 25). (8.15)
7. Cf. *nanda* (9) *paṅktibhir* (10) *athendriyā* (5) *ṃśakaiḥ*. (22.57)
8. See *Śilparatna* of Śrīkumāra :

tadadho munibhiḥ sārdham guṇāṃśaiḥ stambhakuṭṭimam/
dharma-rudrā-rka-bhāgais tu satrayodaśamārgikaiḥ//
aṣṭabhaumam vadanty asminn evam prājñā munīśvarāḥ//

Here the numerals used are *muni* (7), *guṇa* (3), *dharma* (6), *rudra* (11) and *arka* (12).

9. See *Indian Antiquary*, 21 (1902) 48.
10. E.g. Inscription dates *Saṃ* 898 of Caṇḍamahāsena of the Chāhamāna dynasty from Dholpur which reads: *vasunava aṣṭau varṣagatasya kālasya vikramākhyasya*; and copperplate of a gift-deed dated Śaka 867 of the Eastern Cālukya King Amma I: *giri-rasa-vasu-saṃkhyābde śakasamaye*.
11. For instance *Haribhāskara* closes his work *Padyāmṛtataraṅgiṇī* with the verse:

āṣāḍhamāse suvalakṣapakṣe
'kha-vahni-sapt-endu-'mite śubhebde/
gurau navamyām agamat samāptim
sat-padyapīyūṣataraṅgiṇīyam//

(Mss. No. 15 and 510 in the Khasmchar Collection, City Palace, Jaipur).
The date of the completion of the work *Padyāmṛtataraṅgiṇī* of Haribhāskara as stated by the author is Thursday, in the bright fortnight of the month of *Āṣāḍha* in the year 1730.
For the mention, by a scribe, of the date of his completing the copying of a manuscript, see :

'vahni-vedā-bdhi-candre' ca śālivāhaśakābdike/
navamādhyāyaṭīkeyam likhitā Tañjapattane//

(*Bhāgavata-vyākhyā*, Ms. Tanjore Sarasvati Mahal Library, Cata. No. 9959.)
Here, the year of copying the manuscript given in word-numerals is Śālivāhana-Śaka 1443.

12. See for instance, the directions for the computation of *avama* or omitted lunar days, in the *Mahābhāskarīya* of Bhāskara I :

saṃhatya randhra-yamalaiḥ (29) rasa-rāma (36) bhāḡān
bhūyogni-veda (43) guṇiteṣu harec ca bhāgam/
kha-vyoma-kha-dvi-munibhiḥ (72,000) pralayas tithīnām
saṃyojya bhūta-yama-rudra (1125)-hate dinānì//

(*Mahābhāskarīya* 1.23)

13. A typical instance of the documentation of very large numbers through graceful and sparkling verses might be cited here to illustrate the ease with which authors of astronomical texts handled the word-numerals.

47,23,32,26,54,67,510
vyom-aika-bāṇā-dri-rasā-bdhi-bhūta-
ṣaḍ-dvy-aśvi-rāmā-gni-yamā-dri-vedāḥ/
bhānor, vidhoḥ, khā-bhra-khā-bhrā-bhra-netra-
3,53,30,86,62,00000
ṣaṭ-ṣaḍ-vasu-triṃśad-agnī-ṣu-rāmāḥ//
5,60,22,54,07,11,75,000
khā-bhrā-bhra-bāṇā-dri-śivā-dri-khā-bdhi-
tattvā-kṣi-śūnyā-ṅga-śarā guroś ca/
kakṣyā imās tattva-nagāhiṣaṭka-
16,35,68,725
bhūtā-gni-ṣaḍ-bhūmi-bhir āhatāḥ syuḥ//

(*Parameśvara's* sub-com. *Siddhāntadīpikā* on the *Bhāṣya* of *Govindasvāmin* on the *Mahābhāskarīya*, 1.52, edn., p. 85).

The four numbers documented above have, respectively, fifteen, fourteen, sixteen and nine digits. Incidentally it might be remarked that these mathematicians who could jumble such big numbers with ease and perform mathematical operations with those numbers by means of calculations made on the floor using cowries or using such other simple equipment do deserve kudos even from the mathematicians of today, equipped with sophisticated devices like calculators and computers.

14. Cf. the *Candra-vākyas* of the Kerala astronomer Vararuci, 348 in number, which give the daily angular motion of the Moon in signs, degrees and minutes. Therein, the expression *gīr naḥ śreyaḥ* represents a particular position of the moon on the zodiac as 12^{0}-03'. Besides, it forms also a logical expressioin which means 'Our superiority lies in our words'. The expressions in continuation are *dhenavaḥ śrīḥ* (24^{0}-04') 'cows bring prosperity', *rudras tu namyaḥ* (1^{r}-06^{0}-22'), '*Rudra* is to be revered' etc. These are easily memorised, not liable to corruption

and in case of any doubt can easily be checked for their correctness through their meaning and grammar.

For instances of more detailed tables, see the following from the *Candra-vākyas* composed by *Saṅgamagrāma-Mādhava* (A.D. 1350-1410): *bhāve smaro 'ṅganānāṃ syāt* ($10^r 03^0 25' 44''$), *gāyatrī nāstikair nindyā* ($10^r 16^0 02' 13''$), *dhanaṃ coro haren nityam* ($10^r 28^0 26' 09''$), all sensible and easily remembered expressions and, at the same time, expressing the required angular positions of the moon.

15. Among Sanskrit texts written by modern astronomers might be mentioned: *Gaṇitaprakāśikā* by V.A. Poduval. Kannur, Pub. Desamitram, 1952; *Gaṇitanirṇaya* by Puliyoor Purushottaman Namputiri. Quilon, 2nd edn. 1952; *Śuddha-Dṛggaṇitam* by V.P. Kunjikkanna Poduval, Kozhikode, 1956. All the three works have commentaries in Malayalam written by the authors themselves.
16. This work instructs: *sarvatra savarṇā bhāvā rāśayaś ca. Na grahāḥ* (1.33-34): 'Everywhere *bhāvas* and *rāśis* are indicated (in this work) by letters. But not so the planets' For *sūtras* where Kaṭapayādi notation is used, see I.4-5 etc. (*Jaiminīsūtra*, ed. with tr. by B. Suryanarayana Rao, Bangalore, 1949; ed. with tr. by P.S. Sastri, New Delhi, 1992).
17. S. Madhavan, 'Origin of Kaṭapayādi system of numerals', *Pūrṇatrayī*, Govt. Sanskrit College, Tripunithura (Kerala), 18. ii (July-Dec. 1991), 35-42.
18. *Sūktaślokāḥ*, ed. with Introduction and Notes by C. Kunhan Raja, *Jl. of Oriental Research* (Madras), 8 (1935), 358-71.
19. See, for instance, the undermentioned verse by Nārāyaṇa Bhaṭṭa describing the flood in the river *Bhāratappuzha* in Malabar, each line of the verse depicting the Kali date of the flood, alongside describing the gravity of the flood. The poet has also so contrived the letters that the date is given in the normal way in lines one and three and in the reverse way in lines two and four:

    nadīpuṣṭir asahyā nu, na hy asāraṃ payo 'jani/
    nijāt kuṭīrāt sāyāhne, naṣṭārthāḥ prayayur janāḥ//

    The Kali date referred to is 17,21,180 which fell in July 1611.
20. For the full list see P. Sambamoorthy, The *Melakartā Janyarāga Scheme*, Madras, ed., pp. 14-16; Rangaramanuja Ayyangar, *History of South Indian (Carnatic) Music,* Madras, 1972, pp. 1196-99.
21. Parameśvara states towards the commencement of his work *Dṛggaṇita* (ed. K.V. Sarma, Hoshiarpur, 1963) that computations using the then current *Parahita* system led to the derivation of wrong planetary positions different from what is observed and adds that he was composing his *Dṛggaṇita* according to which computation will tally with observation. Elsewhere he has stated that his present effort was the result of 55 years of sky-watching with the instruments (I.i.2-5).

In his *Siddhāntadīpikā* (*Mahābhāskarīya-Bhāṣya-Vyākhyā*, ed. T.S.K. Sastry, Madras, 1957, p. 332, verses 91-93), he gives the then current rate of correction and adds that in course of time even those corrections shall have to be revised.

22. Nīlakaṇṭha Somayāji declares in his work *Jyotirmīmāṃsā*, 'Investigations on Astronomical Theories' (ed. K.V. Sarma, Hoshiarpur, 1977) that "the *Karaṇa* texts alone have contemporary value and accuracy: *karaṇānām eva hi vyāvahārikatvaṃ sūkṣmatvaṃ ca syāt* (p. 8) and quotes with approbation other texts: *Karaṇa* is the result of numerous experiments '*bahuśo yat parīkṣitam karaṇam*' and 'A *Karaṇa* is one that yields, every day, (computed) results which tally with observation': *dṛṣṭānurūpaṃ karaṇaṃ yad anvaham* (ibid., p. 9). In the same work (pp. 35-36) Nīlakaṇṭha indicates the principle of experimentation to produce results of computation which would be equal to observation. In another of his works entitled *Grahaparīkṣākrama* he dwells on the methodology of astronomical observation with instruments and the way of finding correct results, of course, towards preparing revised *Karaṇa* texts, which will involve depiction through the Kaṭapayādi notation.

23. Jyeṣṭhadeva, author of *Yuktibhāṣā, 'Elucidation of rationales in Indian astronomy'*, (ed., tr. by K.V. Sarma, not published) makes a mention in the post-colophonic verses of his work *Dṛkkaraṇam* about revising current parameters for computation to equal observation and the necessity of correcting them periodically. Interestingly the single palm leaf manuscript of the work (Kerala Univ. Or. Mss. Library, Curator's Collection, Malayalam Sn., No. 7-c) has written, on three leaves at the close, the sine table of the different planets corrected to the month of *Vṛścika*, Kollam era 1047, corresponding to A.D. 1872, October, in Kaṭapayādi notation.

24. For Mss. catalogued under the titles '*Jyāḥ*' and '*Jyotiṣavākyāni*' in the Kerala Univ. Or. Mss. Library, Trivandrum, see its Alphabetical List of Sanskrit Mss., Nos. 5988-99, 6024-36, 25430-34 and 25442-58 in Vols. I and V. See also Index of Malayalam Mss., Nos. 3539-59, 3593.

25. See pp. 135-249, being Appendix III to *Vākyakaraṇa* with the commentary of Sundararāja (eds. T.S.K. Sastry and K.V. Sarma, K.S.R. Institute, Madras, 1962). The *Vākyas* commence as follows: *Kujasya Mahāvākyāni: Parivṛttiḥ* I. *maṅgalaśrīr bhūsūnuḥ* (4^r 23^0 35'), *ātmajayī śantanuḥ* (5^r 18^0 50'), *duṣṭo bhūpatir venaḥ* (6^r 14^0 18') with their *saṃskāra-vākyas* 7, 6 and 4, respectively, indicated by the last letters.

26. Ms. Tanjore Sarasvati Mahal Library, No. 11430. The last three mnemonics are: *dhanyākṣī bālā* (3^0 36' 19"), *kamo 'ram bhūmipānām* (15^0 42' 51") and *syād balavān sukhī* (27^0 43' 31").

27. On this, see the commentary of Sūryadeva Yajvan on *Āryabhaṭīya* 1, 2 (ed. K.V. Sarma INSA, New Delhi, 1976, p. 10).
28. Parameśvara of Vaṭaśreṇī in Kerala states the system in a simplified manner:

    kādi-makārāntaiḥ syuḥ saṅkhyā ekādikās tu vargasthāḥ/
    yādir avarga-saṁsthāḥ triṁśatpūrvā daśaka-saṅkhyāḥ//
    kādi-yakārādisthaiḥ svarair avarṇādi-saṃjñayā kathitaiḥ/
    vedyaṃ prathamādipadaṃ vargākhyeṣu kramāt, tathānyeṣu//

    (Com. *Siddhāntadīpikā* on Govindasvāmin's *Bhāṣya* on Bhāskara I's *Mahābhāskarīya* 1.8 (ed. T.S. K. Sastry, GOML, Madras, pp. 13-14).
29. *Mahāsiddhānta*, ed. by Sudhakara Dvivedi with his own commentary, Banares, 1910; The *Pūrvagaṇita* of *Āryabhaṭa's* (II) *Mahāsiddhānta*, ed. and tr. by S. R. Sarma. 2 Pts., Marburg, 1966.
30. See, for instance, the depiction of the planetary revolutions:

    kalpe sūryādīnāṃ bhāgaṇā ghaḍaphenanenanunīnāḥ/
    mathathamagaglabhananunā khakhajhatajogīpanīnonāḥ//
    kasadhagasanamaghacasipā bocībhorīkudhitahīrāḥ/
    sīnaragasakaghaḍaṭhā kaḍhatīmatīmotīdhanīnenāḥ//

    (*Mahāsiddhānta*, 1.7-8)

    Contrast the above-said wording of the numbers with the same numbers given by Nīlakaṇṭha Somayāji in his *Tantrasaṅgraha* in word-numerals for their grace and fluency:

    divyābdānāṃ sahasrāṇi dvādaśaikaṃ caturyugam/
    sūryasya paryāyas, tasmād ayutaghnaradārṇavāḥ/
    khāśvideveṣusaptādriśarāś cendoḥ, kujasya tu/
    vedāṅgāhirasāṅkāśvikarā, jñasya khaparyayāḥ//
    nāgavedanabhaḥsaptarāmāṅkasvarabhūmayaḥ/
    vyomāṣṭarūpavedāṅgapākāś ca bṛhaspateḥ//
    aṣṭāṅgadasranetrāśvikhādrayo bhṛguparyayāḥ/
    bhāskarāṅgarasendrāś ca śaneḥ, śaśyuccapātayoḥ/
    netrārkāṣṭāhivedaś ca, khakharāmaradāśvinaḥ//

    (*Tantrasaṅgraha* of Nīlakaṇṭha Somayāji, 1.15-19a)

    Compare also the yuga revolutions set out in Kaṭapayādi-I :
    nānājñanapragalbnaḥ, tilabalmamasusūkṣmam, dhayed rājadambho
    bhadrodanto dharendro, niranusṛgadhisaukhyam, variṣṭho 'bhiṣaṅgaḥ//
    dordaṇḍāgredrināṭho, viṣamitavipinam, candrarekhāmbukhinnety
    arkādeḥ paryayāḥ syuḥ, kṣitidinam anṛśaṃsaḥ kaḷārthī sa martyaḥ//

    (*Karaṇapaddhati* of Putumana Somayāji, 1.3).

31. See M.K. Gurukkal, *Kalāvidyā-vivaraṇam* (in Malayalam), Kozhikode, 1937, pp. 179-80.
32. *Bhāratīya Prācīna Lipimālā: The Palaeography of India* (in Hindi), Ajmer, 1918, pp. 107-8. See also B.B. Dutta and A.N. Singh, *History of Hindu Mathematics*, Bombay, 1962, pp. 72-74.
33. See *Prācīna-Lipimālā*, *op. cit.*, p. 7; Burnell, *South Indian Palaeography*, London, 1878, p. 79.
34. *Prācīna-Lipimālā*, *op. cit.*, p. 124; Burnell, *op. cit.*, p. 79.
35. *Prācīna-Lipimālā*, *op. cit.*, p. 124; Burnell, *op. cit.*, p. 79.

APPENDIX I

Bhūtasaṅkhyā System[1]
Word-Numerals in Sanskrit

A. WORDS FOR NUMBERS

0. *ananta, antarikṣa, abhra, ambara, ambuda, asat, ākāśa, khaṃ, kha, gagana, chidra, jalada, jaladhara, jaladharapatha, diva, nagna, nabha, payoda, payodhara, puṣkara, pūrṇa, pṛthu, pṛthula, bindu, megha, randhra, varīya, vindu, viyat, viṣṇupada, viṣṇupāda, vihāyas, vṛhat, vyoma, śūnya.*

1. *aṃśumān, acalā, abja, amṛtāṃśu, avani, ādi, āsya, indu, ilā, uḍupati, urvarā, urvī, ṛkṣeśa, eka, eṇadhara, auṣadhīśa, ka, kalādhara, kali, ku, kumudākarapriya, kṣapākara, kṣamā, kṣiti, kṣoṇi, kṣoṇī, kṣmā, go, gotra, gotrā, glau, candra, candramā, jagatī, jaivātṛka, jyā, tanu, dākṣāyaṇīprāṇeśa, dharaṇī, dharā, dharitrī, nāyaka, niśākara, niśeśa, pitāmaha, pṛthivī, pṛthvī, prāleyāṃśu, brahma, bhuvanyu, bhū, bhūmi, mahī, mukha, mṛgalāñchana, mṛgāṅka, medinī, rajanīkara, rajanīśa, rātripa, rātrīśa, rugṇa, rūpa, lapana, vaktra, vadana, vasudhā, vasundharā, vāk, vidhu, viriñci, viśvambharā, śaśadhara, śaśabhṛt, śaśalāñchana, śaśāṅka, śaśi, śaśī, śītakara, śītakiraṇa, śītadyuti, śītamayūkha, śītaraśmi, śītāṃśu, śubhrabhānu, śveta, śvetāṃśu, śītaraśmi, sudhāṃśu, soma, sthirā, hariṇadhṛt, hariṇāṅka, himakara, himagu, himaraśmi, himāṃśu.*

2. *aṃsa, akṣi, antaka, ambaka, ayana, aśvi, aśvinī, īkṣaṇa, oṣṭha, kara, karṇa, kuca, kuṭumba, garut, gulpha, cakṣu, jaṅghā, jānu, dasra, dṛś, dṛṣṭi, doḥ, dvandva, dvaya, dvi, dhruva, naya, nayana, nāsatyau, netra, pakṣa, pāṇi, bāhu, bhuja, bhujā, yama, yamala, yuga, yugala, yugma, ravicandrau, raviputra, locana, śrotra, stana, hasta.*

3. *agni, anala, ājyabjuk, ājyāśa, uṣarbudha, kāla, kṛśānu, kṛṣṇavartma, krama, guṇa, gupti, jagat, jyoti, jvalana, jvāla, tapana, tṛtīya, traya, trikaṭu, trikāla, trigata, triguṇa, trijagat, trinetra, tripadī, traita, daha, dahana, dīpti, dyuti, dhāma, netra, pada, pāvaka, puṃ, pura, puruṣa, puṣkara, brahman, bhuvana, ratna, rāma, rudrākṣa, rohita, loka, locana, vacana, vahni, vikrama, viṣṇu, viṣṇukrama, vṛtta, vaiśvānara, śakti, śikhā, śikhin, śivanetra, śūla, saptārci, sahodara, haranayana, haranetra, havyabhuk, havyavāhana, havyāśa, hutabhuj, hutavaha, hutāśa, hutāśana, hotṛ.*

4. *apas, abdhi, ambu, ambudhi, ambunidhi, ambhodhi, ambhonidhi, aya, arṇava, āpa, āmnāya, āya, āśrama, udadhi, udanvān, upāya, kaṣāya, kṛta, kendra, koṣṭha, gati, jala, jaladhi, jalanidhi, turīya, turya, toya, dik, diś, diśā, dhyāna,*

nadī, nigama, nimnageśa, nīradhi, nīranidhi, payodhi, payonidhi, pātha, pāthodhi, pitāmahāsya, praṇimnageśa, bandha, bandhu, brahmodbhava, mahormi, mahormika, māla, yuga, lavaṇoda, lokapāla, varṇa, vāridhi, vārinidhi, vārdhi, viṣadhi, viṣanidhi, veda, śruti, sañjñā, samudra, salilākara, sāgara, sukha.

5. *akṣa, anila, anuttara, artha, asu, āśuga, indriya, iṣu, kaṅkapatrī, karaṇīya, kalaṅka, kāmaguṇa, kāmadeva, khaga, tattva, tanmātra, dhī, nārāca, nālīka, pakṣi, pañca, pataṅga, patrin, parva, pavana, pāṇḍava, putra, pṛṣatka, prāṇa, bāṇa, bhāva, bhūta, bhedaka, marut, mahābhūta, mahāvrata, mārgaṇa, ratna, leya, leyaka, vadha, varṣman, vāta, vāyu, viśikha, viṣaya, vrata, śara, śarīra, śastra, śilīmukha, śuka, śvasana, samiti, sāyaka, smarabāṇa, smareṣu.*

6. *aṅga, abda, arāti, ari, ṛtu, kāya, kāraka, kārtika, kāla, kumāra, kumāravadana, kṛttikā, koṣa, kṣmākhaṇḍa, khara, guhaka, guhavaktra, guhāsya, cakrārdha, camū, caraka, jīva, tarka, darśana, dravya, bhārdha, māsārdha, rasa, rāga, rāsabha, ripu, lekhya, leśyā, varṇa, śakti, śāstra, śilīmukhapada, śiśu, ṣaṭ, ṣaṇmukha, samāsa, senā, skanda, svāda.*

7. *aga, acala, adri, anila, aśva, ṛṣi, kalatra, kulācala, kṣitibhṛt, kṣmādhāra, kṣmābhṛt, gandhanvaha, gandhavaha, giri, ghoṭā, chanda (chandas), tattva, tāpasa, turaga, turaṅga, divicara, dvīpa, dharaṇidhara, dharādhara, dharitrīdhara, dhātu, dhī, naga, pannaga, parvata, pavana, bhaya, bhūdhara, bhūdhra, bhūbhṛt, marut, mahīdhara, mahīdhra, mātṛkā, māruta, muni, yati, loka, vakrī, vāji, vāta, vāyu, vāra, vāsara, vyasana, śilā, śiloccaya, śaila, sapta, samīra, samīraṇa, sthira, svara, haya.*

8. *anīka, anuṣṭubh, ahi, ibha, uraga, karaṭin, karī (karin), kareṇu, karma (karman), kuñjara, gaja, gahana, gahvara, takṣa, tanu, dantī (dantin), dik, diggaja, diṅnātha, diś, durita, dvipa, dvirada, dhī, dhīguṇa, nāga, nāgendra, padmī, pannaga, pavanabhuk, pīluka, puṣkarin, pradhāna, prabhāvaka, phaṇin, buddhiguṇa, bhadra, bhujaga, bhujaṅga, bhujaṅgama, bhūti, maṅgala, mataṅgaja, mati, mada, mātaṅga, yūthapa, vana, vasu, vāraṇa, viṣabhṛdvapu, vyāla, sriya, sarpa, siddhaguṇa, siddhi, sindhura, suṇḍipa, stamberama, stambherama, hastin.*

9. *aṅka, antara, ambujāsana, upendra, ṛddhi, kapāṭa, kavāṭa, kavi, kūpa, keśava, kṣata, khaga, khacara, khegāmī, khecara, gambhīra, gāmbhīra, go, graha, chidra, tāmisra, tārkṣyadhvaja, divisad, durgā, dvāra, nanda, nabhoga, nava, nidhāna, nidhi, paṭa, padārtha, bala, bila, miti, yantra, ratna, randhra, labdha, labdhi, vaṭa, vila, vivara, śuṣira, suṣira, hari.*

10. *aṅguli, aṅgulī, avatāra, āśā, kakubh, karma, kāṣṭha, kāṣṭhā, dik, dikpāla, diś, diśā, paṅkti, rāvaṇaśira, harit.*

11. *akṣauhiṇī, aja, īśa, īśāna, īśvara, giriśa, girīśa, guṇī, gaurīśa, tryambaka, dhūrjaṭi, paśupati, pinākin, bharga, bhava, mahaḥ, mahādeva, maheśa, maheśvara,*

mṛḍa, rudra, śaṅkara, śambhu, śarva, śaśibhūṣaṇa, śitikaṇṭha, śiva, śūlin, śrīkaṇṭha, hara.

12. *arka, āditya, ina, uṣṇāṃśu, gaṇa, cakra, tapana, taraṇi, tigmāṃśu, tīkṣṇāṃśu, dinakara, dinanātha, dinapa, dinamaṇi, dineśa, divākara, dyumaṇi, nara, nā, pataṅga, pūṣā, pradyotana, bhagaṇa, bhānu, bhānumān, bhāskara, maṇḍala, mārtaṇḍa, māsa, mitra, mihira, ravi, rāśi, vivasvān, vyaya, śaṅku, savitā, sūrya, haṃsa.*

13. *aghoṣa, atijagatī, anaṅga, karaṇa, kāma, kriyāsthāna, jaya, yakṣa, vandya, viśva, viśve, viśvedevāḥ.*

14. *indra, indraka, kauśika, guṇasthāna, jiṣṇu, purandara, pūrvapa, bhuvana, bhūtagrāma, manu, rajju, loka, vajrin, vāsava, vidyā, śakra, śakvarī, śacīpati, surādhipa, sureśa.*

15. *atiśakvarī, ahan, ahnisaṅkhyā, ghasra, tithi, dina, divasa, dyuniśā, pakṣa, paramādhārmika, vāsara, sandhyā.*

16. *avanipa, aṣṭi, indukalā, kalā, kṣmāpa, nṛpa, nṛpati, pṛthvīśa, bhūpa, bhūpati, bhūmipa, ritha, śaśikalā.*

17. *atyaṣṭi, ghaṭā, ghana, jīmūta, toyada, balāha, balāhaka, megha, maitra, valāha, valāhaka, vidhṛti.*

18. *abrahma, tīrtha, dhṛti, pāpasthānaka, purāṇa, vidyā, smṛti.*

19. *atidhṛti, vagha, sāma.*

20. *aṅgula, karaja, kṛti, nakha, nakhara.*

21. *kratu, prakṛti, bṛnda, makha, mūrccha, mūrcchanā, yajña, vṛnda, samidhaḥ, svaḥ, svarga.*

22. *ākṛti, ākṛtī, jāti, jātī, parīṣaha, mati.*

23. *aja, jahu, māyā, meṣa, vikṛti.*

24. *andha, arhat, gāyatrī, jina, tīrthaka, tīrthakṛt, dhamma (dharma), buddha, saṃkṛti, siddha.*

25. *atikṛti, abhikṛti, jñāna, tattva, puṣṭa.*

26. *atitattva, utkṛti.*

27. *uḍu, ṛkṣa, khadyota, tāra, tāraka, tārakā, tārā, dhiṣṇya, nakṣatra, bha, bhāni.*

28. *ṛkṣa, deha, piṇḍa, hulā.*

29. *āma, uṣṭi, kuva.*

30. *jaṭā, tithi, dina, sādhvī.*

32. *danta, daśana, rada, radana.*

33. *amara, asvapna, gīrvāṇa, tridaśa, deva, daivata, nākagha, nirjara, sura.*

39. *dūrvā.*

40. *tāla, paṅkti.*

45. *tripakṣī.*
48. *jagatī, saṃskāra.*
49. *anila, tāna, pavana, marut, vāyu.*
55. *suṣṭi.*
59. *kāca, mokṣa.*
64. *kalā, yoginī, śrīkalā.*
65. *puṣṭi.*
72. *puruṣakalā.*
360 *cakrāṃśa, bhāṃśa.*
5400. *darada.*
21600. *cakrakalā, cakraliptā.*

B. NUMBERS FOR WORDS

Arranged according to the Sanskrit alphabet.

aṃśumān 1
aṃśa 2
akṣa 5
akṣi 2
akṣauhiṇī 11
aga 7
agni 3
aghoṣa 13
aṅka 9
aṅga 6
aṅgula 20
aṅguli or *aṅgulī* 10
acala 7
acalā 1
aja 11, 23
atikṛti 25
atijagatī 13
atitattva 26
atidhṛti 19
atiśakvarī 15
atyaṣṭi 17
adri 7
anaṅga 13
ananta 0
anala 3
anila 5, 7, 49
anīka 8
anuttara 5
anuṣṭubh 8
antaka 2
antara 9
antarikṣa 0
andha 24
apas 4
abja 1
abda 6
abdhi 4
abrahma 18
abhikṛti 25
abhra 0
amara 33
amṛtāṃśu 1
ambaka 2
ambara 0
ambu 4
ambujāsana 9
ambuda 0
ambudhi 4
ambunidhi 4
ambodhi 4
ambhonidhi 4
aya 4
ayana 2
arāti 6
ari 6
arka 12
arṇava 4
artha 5
arhat 24
avatāra 10
avani 1
avanipa 16
aśva 7
aśvi 2
aśivinī 2
aṣṭi 16
asat 0
asu 5
asvapna 33

ahan 15
ahi 8
ahnisaṃkhyā 15
ākāśa 0
ākṛti 22
ākṛtī 22
ājyabhuk 3
ājyāśa 3
ādi 1
āditya 12
āpa 4
āma 29
āmnāya 4
āya 4
āśā 4
āśuga 5
āśrama 4
āsya 1
ina 12
indu 1
indukalā 16
indra 14
indraka 14
indriya 5
ibha 8
ilā 1
iṣu 5
īkṣaṇa 2
īśa 11
īśāna 11
īśvara 11
uḍu 27
uḍupati 1
utkṛti 26
udadhi 4
udanvān 4
upāya 4
upendra 9
uraga 8
urvarā 1
urvī 1
uṣarbudha 3
uṣṭi 29
uṣṇāṃśu 12
ṛkṣa 27, 28
ṛkṣeśa 1
ṛtu 6
ṛddhi 9
ṛṣi 7
eka 1
eṇadhara 1
oṣṭha 2
auṣadhīśa 1
ka 1
kakubh 10
kaṅkapatrī 5
kapāṭa 9
kara 9
karaja 20
karaṭin 8
karaṇa 13
karaṇīya 5
karī (karin) 8
kareṇu 8
karṇa 2
karma (karman) 8, 10
kalaṅka 5
kalatra 7
kalā 16, 64
kalādhara 1
kali 1
kavāṭa 9
kavi 9
kaṣāya 4
kāca 59
kāma 13
kāmaguṇa 5
kāmadeva 5
kāya 6
kāraka 6
kārtika 6
kāla 3, 6
kāṣṭha 10
kāṣṭhā 10
ku 1
kuca 2
kuñjara 8
kuṭumba 2
kumāra 6
kumāravadana 6
kumudākarapriya 1
kulācala 7
kuva 29
kūpa 9
kṛta 4
kṛti 20
kṛttikā 6
kṛśānu 3
kṛṣṇavartma 3
kendra 4
keśava 9
koṣa 6
koṣṭha 4
kauśika 14
kratu 21
krama 3
kriyāsthāna 13
kṣata 9
kṣapākara 1
kṣamā 1
kṣiti 1
kṣitibhṛt 7
kṣoṇi 1
kṣoṇī 1

kṣmā 1
kṣmākhaṇḍa 6
kṣmādhara 7
kṣmāpa 16
kṣmābhṛt 7
khaṃ 0
kha 0
khaga 5, 9
khacara 9
khaddyota 27
khara 6
khegāmī 9
khecara 9
gagana 0
gaja 8
gaṇa (= bhagaṇa) 12
gati 4
gandhanvaha 7
gandhavaha 7
gambhīra 9
garut 2
gahana 8
gahvara 8
gāmbhīra 9
gāyatrī 25
giri 7
giriśa 11
girīśa 11
gīrvāṇa 33
guṇa 3
guṇasthāna 14
guṇī 11
gupti 3
gulpha 2
guhaka 6
guhavaktra 6
guhāsya 6
go 1, 9
gotra 1
gotrā 1
gaurīśa 11
glau 1
graha 9
ghaṭā 17
ghana 17
ghasra 15
ghoṭā 7
cakra 12
cakrakalā 21600
cakraliptā 21600
cakrāṃśa 360
cakrārdha 6
cakṣu 2
candra 1
candramā 1
camū 6
caraka 6
chanda (chandas) 7
chidra 0, 9
jagat 3
jagatī 1, 48
jaṅghā 2
jaṭā 30
jaya 13
jala 4
jalada 0
jaladhara 0
jaladharapatha 0
jaladhi 4
jalanidhi 4
jahu 23
jāti 22
jātī 22
jānu 2
jina 24
jiṣṇu 14
jīmūta 17
jīva 6
jaivātṛka 1
jyā 1
jyoti 3
jñāna 25
jvalana 3
jvāla 3
takṣa 8
tattva 5, 7, 25
tanu 1, 8
tanmātra 5
tapana 3, 12
taraṇi 12
tarka 6
tāna 49
tāpasa 7
tāmisra 9
tāra or *tāraka* 27
tārakā 27
tārā 27
tārkṣyadhvaja 9
tāla 40
tigmāṃśu 12
tithi 15, 30
tīkṣṇāṃśu 12
tīrtha 18
tīrthaka 24
tīrthakṛt 24
turaga 7
turaṅga 7
turīya 4
turya 4
tṛtīya 3
toya 4
toyada 17
traya 3
trikaṭu 3

trikāla 3
trigata 3
triguṇa 3
trijagat 3
tridaśa 33
trinetra 3
tripakṣī 45
tripadi 3
traita 3
tryambaka 11
danta 32
dantī (dantin) 8
darada 5400
darśana 6
daśana 32
dasra 2
daha 3
dahana 3
dākṣāyaṇīprāṇeśa 1
dik 4, 8, 10
dikpāla 10
diggaja 8
diṅnātha 8
dina 15, 30
dinakara 12
dinanātha 12
dinapa 12
dinamaṇi 12
dineśa 12
diva 0
divasa 15
divākara 12
divicara 7
divisad 9
diś 4, 8, 10
diśā 4, 8, 10
dīpti 3
durita 8
durgā 9
dūrvā 39
dṛś 2
dṛṣṭi 2
deva 33
deha 28
daivata 33
doḥ 2
dyuti 3
dyuniśā 15
dyumaṇi 12
dravya 6
dvandva 2
dvaya 2
dvāra 9
dvi 2
dvipa 8
dvirada 8
dvīpa 7
dharaṇidhara 7
dharaṇī 1
dharā 1
dharādhara 7
dharitrī 1
dharitrīdhara 7
dhamma (dharma) 24
dhātu 7
dhāma 3
dhiṣṇya 27
dhī 5, 7, 8
dhīguṇa 8
dhūrjaṭi 11
dhṛti 18
dhyāna 4
dhruva 2
nakṣatra 27
nakha 20
nakhara 20
naga 7
nagna 0
nadī 4
nanda 9
nabha (nabhas) 0
nabhoga 9
naya 2
nayana 2
nara 12
nava 9
nā 12
nākagṛha 33
nāga 8
nāgendra 8
nāyaka 1
nārāca 5
nālīka 5
nāsatyau 2
nigama 4
nidhāna 9
nidhi 9
nimnegeśa 4
nirjara 33
niśākara 1
niśeśa 1
nīradhi 4
nīranidhi 4
nṛpa 16
nṛpati 16
netra 2, 3
pakṣa 2, 15
pakṣi 5
paṅkti 10, 40
pañca 5
paṭa 9
pataṅga 5, 12
patrin 5
pada 3

padārtha 9
padmī 8
pannaga 7, 8
payoda 0
payodhara 0
payodhi 4
payonidhi 4
paramādhārmika 15
parīṣaha 22
parva 5
parvata 7
pavana 5, 7, 49
pavanabhuk 8
paśupati 11
pāṇi 2
pāṇḍava 5
pātha 4
pāthodhi 4
pāpasthānaka 18
pāvaka 3
piṇḍa 28
pitāmaha 1
pitāmahāsya 4
pinākin 11
pīluka 8
puṃ 3
putra 5
pura 3
purandara 14
purāṇa 18
puruṣa 3
puruṣakalā 72
puṣkara 0, 3
puṣkarin (= hastin) 8
puṣṭa 25
puṣṭi 65
pūrṇa 0
pūrvapa 14
pūṣā (pūṣan) 12
pṛthivī 1
pṛthu 0
pṛthula 0
pṛthvī 1
pṛthvīśa 16
pṛṣatka 5
prakṛti 21
praṇimnageśa 4
pradyotana 12
pradhāna 8
prabhāvaka 8
prāṇa 5
prāleyāṃśu 1
phaṇin 8
bandha 4
bandhu 4
bala 9
balāha 17
balāhaka 17
bāṇa 5
bāhu 2
bindu 0
bila 9
buddha 24
buddhiguṇa 8
bṛnda 21
brahman 3
brahmā 1
brahmodbhava 4
bha 27
bhagaṇa 12
bhadra 8
bhaya 7
bharga 11
bhava 11
bhāṃśa 360
bhāni 27
bhānu 12
bhānumān 12
bhārdha 6
bhāva 5
bhāskara 12
bhuja 2
bhujaga 8
bhujaṅga 8
bhujaṅgama 8
bhujā 2
bhuvana 3, 14
bhuvanyu 1
bhū 1
bhūta 5
bhūtagrāma 14
bhūti 8
bhūdhara 7
bhūdhra 7
bhūpa 16
bhūpati 16
bhūbhṛt 7
bhūmi 1
bhūmipa 16
bhedaka 5
makha 21
maṅgala 8
maṇḍala 12
mataṅgaja 8
mati 8, 22
mada 8
manu 14
marut 5, 7, 49
mahaḥ 11
mahādeva 11
mahābhūta 5
mahāvrata 5
mahī 1
mahīdhara 7

mahīdhra 7
maheśa 11
maheśvara 11
mahormi 4
mahormika 4
mātaṅga 8
mātṛkā 7
māyā 23
māruta 7
mārgaṇa 5
mārtāṇḍa 12
māla 4
māsa 12
māsārdha 6
miti 9
mitra 12
mihira 12
mukha 1
muni 7
mūrcchā 21
mūrcchanā 21
mṛgalāñchana 1
mṛgāṅka 1
mṛḍa 11
megha 17
medinī 1
meṣa 23
maitra 17
mokṣa 59
yakṣa 13
yajña 21
yati 7
yantra 9
yama 2
yamala 2
yuga 2, 4
yugala 2
yugma 2
yūthapa 8
yoginī 64
rajanīkara 1
rajanīśa 1
rajju 14
ratna 3, 5, 9
rada 32
radana 32
randhra 0, 9
ravi 12
ravicandrau 2
raviputra (= yama) 2
rasa 6
rāga 6
rātripa 1
rātrīśa 1
rāma 3
rāvaṇaśira 10
rāśi 12
rāsabha 6
ritha 16
ripu 6
rugṇa 1
rudra 11
rudrākṣa 3
rūpa 1
rohita (= rohitāśva) 3
lapana 1
labdha 9
labdhi 9
lavaṇoda 4
lekhya 6
leya 5
leyaka 5
leśyā 6
loka 3, 7, 14
lokapāla 4
locana 2, 3
vaktra 1
vakrī 7
vacana 3
vajrin 14
vaṭa 9
vadana 1
vadha 5
vana 8
vandya 13
varīya 0
varṇa 4, 6
varṣman 5
valāha 17
valāhaka 17
vasu 8
vasudhā 1
vasundharā 1
vahni 3
vāk 1
vāji 7
vāta 5, 7, 49
vāyu 5, 7, 49
vāra 7
vāraṇa 8
vāridhi 4
vārinidhi 4
vāsara 7, 15
vāsava 14
vikṛti 23
vikrama 3
vidyā 14, 18
vidhu 1
vidhṛti 17
vindu 0
viyat 0
viriñci 1
vila 9
vivara 9

vivasvān 12
viśikha 5
viśva 13
viśvambharā 1
viśve 13
viśvedevāḥ 13
viṣadhi 4
viṣanidhi 4
viṣabhṛdvapu 8
viṣaya 5
viṣṇu 3
viṣṇukrama 3
viṣṇupada 0
viṣṇupāda 0
vihāyas 0
vṛtta 3
vṛnda 21
vṛhat 0
veda 4
vaiśvānara 3
vyaya 12
vyasana 7
vyāla 8
vyoma 0
vrata 5
śakti 3, 6
śakra 14
śakvarī 14
śaṅkara 11
śaṅku 12
śacīpati 14
śambhu 11
śara 5
śarīra 5
śarva 11
śaśadhara 1
śaśabhṛt 1
śaśalāñchana 1
śaśāṅka 1
śaśi 1
śaśī 1
śaśikalā 16
śaśibhūṣaṇa 11
śastra 5
śāstra 6
śikhā 3
śikhin 3
śitikaṇṭha 11
śilā 7
śilīmukha 5
śilīmukhapada 6
śiloccaya 7
śiva 11
śivanetra 3
śiśu 6
śītakara 1
śītakiraṇa 1
śītadyuti 1
śītamayūkha 1
śītaraśmi 1
śītāṃśu 1
śubhrabhānu 1
śuṣira 9
śūka 5
śūnya 0
śūla 3
śūlin 11
śaila 7
śrī 8
śrīkaṇṭha 11
śruti 4
śrotra 2
śvasana 5
śveta 1
śvetāṃśu 1
ṣaṭ 6
ṣaṇmukha 6
saṃskāra 48
saṅkṛti 24
sañjña 4
sandhyā 15
sapta 7
saptārci 3
samāsa 6
samiti 5
samidhaḥ 21
samīra 5, 7, 49
samīraṇa 5, 7, 49
samudra 4
sarpa 8
salilākara 4
savitā 12
sahodara 3
sāgara 4, 7
sādhvī 30
sāma 19
sāyaka 5
sitaraśmi 1
siddha 24
siddhaguṇa 8
siddhi 8
sindhura 8
sukha 4
suṇḍipa 8
sudhāṃśu 1
sura 33
surādhipa 14
sureśa 14
suṣira 9
suṣṭi 55
sūrya 12
senā 6
soma 1
skanda 6

stana 2
stamberama 8
stambherama 8
strīkalā 64
sthira 7
sthirā 1
smarabāṇa 5
smareṣu 5
smṛti 18
svaḥ 21
svara 7
svarga 21
svāda 6
haṃsa 12
haya 7
hara 11
haranayana 3
haranetra 3
hari 9
hariṇadhṛt 1
hariṇāṅka 1
harit 10
havyabhuk 3
havyavāhana 3
havyāśa 3
hasta 2
hastin 8
himakara 1
himagu 1
himaraśmi 1
himāṃśu 1
hutabhuj 3
hutavaha 3
hutāśa 3
hutāśana 3
hulā 28
hotṛ 3

APPENDIX II[1]

[Numbering of the Pages of a Pali Manuscript]

The numbering of the pages of a manuscript is done on each leaf and not on each page. The two pages of a leaf have only one mark, placed on the first page.

The first leaf of a book is marked with the letter *ka*, the second with *kā* etc. The 34 consonants of the Sanskrit alphabet, combining with the 14 vowels, the *anusvāra* and the *visarga,* make a total of 544 signs. A consonant with the *anusvāra* is written as *kāṃ*, *khāṃ etc*. Where the number of leaves in a book exceeds 544, as in the *Samantapāsādikā* and the *Jātakaṭṭakathā* of this collection, the numbering recommences with *ka* but with the number 2 in front. Thus, for instance, 2 *kū* denotes 544 + 6 = 550 and 2 *ṅo* denotes 544 + 77 = 621.

A table giving full details for the deciphering of the number of a leaf is given below for ready reference.

1. From Adikaram, E.W., *Descriptive Catalogue of the Pali Manusctipts in the Adyar Library*, The Adyar Library, Madras, 1947, pp. xv-xvi.

NUMBERING OF THE PAGES OF A MANUSCRIPT

	a	*ā*	*i*	*ī*	*u*	*ū*	*ṛ*	*ṝ*	*ḷ*	*ḹ*	*e*	*ai*	*o*	*au*	*aṃ*	*aḥ*
	ka	*kā*	*ki*	*kī*	*ku*	*kū*	*kṛ*	*kṝ*	*kḷ*	*kḹ*	*ke*	*kai*	*ko*	*kau*	*kāṃ*	*kaḥ*
	1	2	3	4	5	6	7	8	9	10	11	12	13	14	15	16
kha	17	18	19	20	21	22	23	24	25	26	27	28	29	30	31	32
ga	33	34	35	36	37	38	39	40	41	42	43	44	45	46	47	48
gha	49	50	51	52	53	54	55	56	57	58	59	60	61	62	63	64
ṅa	65	66	67	68	69	70	71	72	73	74	75	76	77	78	79	80
ca	81	82	83	84	85	86	87	88	89	90	91	92	93	94	95	96
cha	97	98	99	100	101	102	103	104	105	106	107	108	109	110	111	112
ja	113	114	115	116	117	118	119	120	121	122	123	124	125	126	127	128
jha	129	130	131	132	133	134	135	136	137	138	139	140	141	142	143	144
ña	145	146	147	148	149	150	151	152	153	154	155	156	157	158	159	160
ṭa	161	162	163	164	165	166	167	168	169	170	171	172	173	174	175	176
ṭha	177	178	179	180	181	182	183	184	185	186	187	188	189	190	191	192

ḍa	193	194	195	196	197	198	199	200	201	202	203	204	205	206	207	208
ḍha	209	210	211	212	213	214	215	216	217	218	219	220	221	222	223	224
ṇa	225	226	227	228	229	230	231	232	233	234	235	236	237	238	239	240
ta	241	242	243	244	245	246	247	248	249	250	251	252	253	254	255	256
tha	257	258	259	260	261	262	263	264	265	266	267	268	269	270	271	272
da	273	274	275	276	277	278	279	280	281	282	283	284	285	286	287	288
dha	289	290	291	292	293	294	295	296	297	298	299	300	301	302	303	304
na	305	306	307	308	309	310	311	312	313	314	315	316	317	318	319	320
pa	321	322	323	324	325	326	327	328	329	330	331	332	333	334	335	336
pha	337	338	339	340	341	342	343	344	345	346	347	348	349	350	351	352
ba	353	354	355	356	357	358	359	360	361	362	363	364	365	366	367	368
bha	369	370	371	372	373	374	375	376	377	378	379	380	381	382	383	384
ma	385	386	387	388	389	390	391	392	393	394	395	396	397	398	399	400
ya	401	402	403	404	405	406	407	408	409	410	411	412	413	414	415	416
ra	417	418	419	420	421	422	423	424	425	426	427	428	429	430	431	432
la	433	434	435	436	437	438	439	440	441	442	443	444	445	446	447	448
va	449	450	451	452	453	454	455	456	457	458	459	460	461	462	463	464
sa	465	466	467	468	469	470	471	472	473	474	475	476	477	478	479	480
śa	481	482	483	484	485	486	487	488	489	490	491	492	493	494	495	496
ṣa	497	498	499	500	501	502	503	504	505	506	507	508	509	510	511	512
ha	513	514	515	516	517	518	519	520	521	522	523	524	525	526	527	528
ḷa	529	530	531	532	533	534	535	536	537	538	539	540	541	542	543	544
		ḷā	*ḷi*	*ḷī*	*ḷu*	*ḷū*	*ḷṛ*	*ḷṝ*	*ḷī*	*ḷī*	*ḷe*	*ḷai*	*ḷo*	*ḷau*	*ḷāṃ*	*ḷaḥ*

Brāhmī Numerals and Decimal Notation : Nature and Evolution

AJAY MITRA SHASTRI

INTRODUCTION

According to the common consensus, the so-called international or Arabic numerals in vogue throughout the globe at present are based on or derived from their Indian predecessors as finalised following the introduction of decimal place-value notation and the zero. It was this fully grown system that was first transmitted to or adopted by the Arabs who played the intermediaries from whom the Europeans received it whereafter in due course it got diffused all over the world. It must be noted, however, that this system represented only the culmination of a long process of anterior evolution, and the earlier numbers[1] are known from early Brāhmī inscriptions[2] which give us their evolving forms and a different numerial notation. In order to distinguish it from the later one based on the concept of zero[3] we may call it ancient Indian numerical notation as against the decimal notation.

ANCIENT SYSTEM

In this ancient system there were in all twenty basic signs as available in early inscriptions: one each for 1 to 9, for tens from 10 to 90 excluding the intermediate mixed numbers, and for 100 and 1000, there being no zero in this sytem. These were only basic signs and it was by appending some strokes or ligatures to their right that the intermediate numbers were expressed.[4] With the help of these twenty basic signs, numbers up to 99999 could be formed.[5] For getting the intermediate numbers from 11 to 99, the signs for 1 to 9 had to be added to the right of the symbols for tens, e.g. 10 (+) 1 = 11, 20 (+) 1 = 21, 90 (+) 9 = 99 etc. For getting 200 and 300 one had to add one and two horizontal strokes (standing for 1 and 2) respectively to the right of the figure for 100, whereas for writing 400 to 900 one was required to attach the figures for 4 to 9 to the right of that for 100 commonly joined together by a short horizontal stroke. Similarly, for forming 2000 and 3000 were to be appended the same signs for 1 and 2 respectively to the right of the figure for 1000,[6] while the figures for 4000 to 9000 could be obtained by appending the signs for 4 to 9 to its right, both connected together by a short horizontal line. The figures for 10000 to 90000 could likewise be formed by suffixing those for 10 to 90 to 1000. Thus, for obtaining 99999 one had to reproduce the signs for 90000, 900, 90 and 9 in that order from left to right.[7]

The oldest visible figures are met with in the edicts of the Maurya emperor Aśoka dating from about early third century BC, but the variety of some of these signs occurring in different versions of the same edict would indubitably show that they were in vogue for a few centuries earlier.[8]

Many of these numeral signs look almost exactly like some Brāhmī letters or their conjuncts. Basing on this closeness in form, some early epigraphists including Bhagwanlal Indraji,[9] H. Kern[10] and G. Bühler[11] held that, as in some other ancient cultures, in India also Brāhmī letters were used as numbers. However, as demonstrated by Gaurishankar Hirachand Ojha after a long critical analysis,[12] the simiarlity is purely accidental and not much weight needs be attached to it. In many cases the development and alteration in the form are not parallel and some of the numerals have no parallel Brāhmī *akṣaras*.[13] Moreover, had the letters actually been intended to represent numbers, some order following the *varṇa-mālā* would have been perceptible which is not the case. There is yet another strong argument against this theory: when some astronomers like Āryabhaṭa I (*b.* 476 A.D.) and his later namesake (Āryabhaṭa II, *c.* 950 A.D.) and their predecessors devised, for brevity's sake, a system of denoting numbers by letters of alphabet, they would have most probably adopted the existing letter-numerals, had any such been obtaining from earlier period and would not have taken trouble to invent an entirely new system; but such is not the case and the letters used for denoting particular numbers are totally unrelated to those that are supposed by these savants to stand for these numbers.[14]

As in the case of the Brāhmī script, attempts have been made to trace the origin of this system also to some western sources, viz. Hieroglyphic, Hieratic or Demotic systems of ancient Egypt.[15] However, there are fundamental differences between the supposed originals and the Brāhmī numerals, demonstrating the utter untenability of these theories. The Hieroglyphic system had only three basic signs, viz. for 1, 10 and 100, and by repeating these symbols one could get numbers up to 300. For expressing higher numbers (1000, 10000, 100000 and 1000000) some strange pictographs were employed. Later, in order to simplify this system twenty signs (9 signs for 1 to 9, 9 signs for tens from 10 to 90 and two for 100 and 1000) as in Indian system were devised. There is absolutely no similarity, it would be noticed, between the Brāhmī and Hieroglyphic systems. Of the Hieratic system also only the sign for 9 is somewhat similar in form to its Brāhmī counterpart. Then, for forming 200, 300 and 400 *one, two and three vertical strokes were added to the left of the sign for 100 as against the Brāhmī system where 200 and 300 were formed by adding one and two horizontal strokes to the right of the sign for 100 and 400 to 900 were indicated by appending the numbers 4 to 9 to the right of 100.* The same is true of the forms of 2000 to 4000 in the Heiratic system which had the figure for 1000 in a somewhat modified form over which 2 to 4 vertical strokes were placed. Then, for 1 to 3 in Brāhmī were employed 1 to 3 horizontal lines placed one above the other, while the Hieroglyphic, Hieratic as well as the latter's modification Demotic employed vertical lines placed side by side, joined together in the last system. It would be obvious from this rather lengthy discussion that there is not much of a similarity between Brāhmī numerical system and its supposed Egyptian originals as well as the forms of the numeral signs.[16]

ZERO AND DECIMAL NOTATION

The complexity and naivety of this system were soon realised, and some unknown genius invented the zero or *śūnya* which made arithmetical calculations extremely simple. Now, with the employment of zero only ten marks, viz. 1 to 9 and 0, could perform even the highest calculations. When the zero came to be invented is at present impossible to determine owing to a total absence of evidence on the point. It has been suggested by some that the word *kṣudra* in the *Atharva-veda* (XIX. 22.6) signifies a trifling or negative number, viz. zero.[17] Whatever that be, the fact remains that it is not found employed in epigraphs for a considerably long time where the old numerals still continue to reign. In inscriptions it is first undoubtedly employed in some records in South-East Asia. At Sambor, Palembang and Banka (Kotakapur) have been discovered inscriptions dated in Śaka years 605, 606 and 608 respectively which are coeval with 683, 684 and 685 A.D.[18] Since the employment of zero in these records composed in Sanskrit and belonging to the Indian immigrants is in the context of the years of the Śaka era of an undoubted Indian origin,[19] there should be no doubt whatever that like the Sanskrit language and the Śaka era, the figure for zero also must have travelled to Cambodia along with the Indians. In view of this Joseph Needham's view that the zero was discovered in South-East Asia including Indo-China and Java where Hindu culture 'met the southern zone of the culture of the Chinese'[20] cannot be conceded. In India proper, however, it is found used in inscriptions of a slightly later date, viz. dating from about eighth-ninth centuries A.D. It occurs in the Ragholī (Bālāghāṭ district, Madhya Pradesh) charter of Śailavaṃśin king Jayavardhana II dating palaeographically from about the eighth century[21] and in one of the documents included in the Kāman (Bharatpur district, Rajasthan) inscription, dated year 220, perhaps of the Harṣa era commencing 606 A.D.[22] In literary sources, however, it is referred to considerably earlier. As we have seen above, some scholars find a reference to the negative or trifling number called *kṣudra* in the *Atharva-veda*. This is, however, doubtful, especially in view of the fact that it is not mentioned in literature for quite a few centuries and not found used, directly or implicitly, up to at least the beginning of the Christian era. There is no doubt, however, about its mention as *śūnya* in the *Chandaḥsūtra* of Piṅgala (400-200 B.C.) in its *saṅkhyā-pratyaya* section in connection with the computation of the total number of arrangement of long (*guru*) and short (*laghu*) syllables without actually constructing the *prastāra* (arrangement of all possible combinations of *guru* and *laghu* [syllables] in a given metre).[23] The *Vāsavadattā* of the celebrated Sanskrit fictionist Subandhu (*c.* fourth century A.D.) compares the stars in a cloudless night to zero-dots (*śūnya-bindavaḥ*), showing thereby that one of the ways of expressing zero was through writing a dot.[24] The Jaina author Siddhasenagaṇi of the sixth century A.D. in his commentary on the *Tattvārthādhigamasūtra* (III.11) makes an interesting observation about the inherent decimal property of *śūnya*.[25] Mention must in this connection be made of a mathematical tract written on birch-barks unearthed at the village of Bakhshālī in the

Yūsufzāī district of Pakistan, commonly called by historians of mathematics as Bakhshāli manuscript. According to A.F.R. Hoernle[26] followed by some other scholars,[27] it dates from early centuries (third/ fourth) A.D.

It not only gives the numbers following the post-zero system but the zero itself which has the form of a dot or *bindu*, a term (*śūnya-bindu*) employed by Subandhu as seen earlier. George Bühler treats Hoernle's 'estimate of the antiquity of the arithmetical treatise' found at Bakshālī as 'very probable' and feels that in such an eventuality zero's 'invention dates from the beginning of our (viz. Christian) era or even earlier.'[28] However, despite these categorical views of such high authorities, it is just impossible to concede such an early date for this text. Written in Śāradā script, the manuscript cannot be placed in its present form earlier than eighth century A.D. or thereabouts.[29] Be that as it may, it has zero in the dot (*bindu*) form. However, in the aforesaid Kambuja inscriptions we notice the forms both of a circle and a dot as at present.[30] And in fact there is no fundamental difference between the two; it depends on individual writing styles and inclinations and the technique applied in writing: if the circle is blackened (in writing) or its interior is left unscooped (in inscription), it would look like a dot and *vice versa*. This should suffice to explain the simultaneous appearance of both the forms, sometimes in the same record.

Mention must in the present context be made of the practice of expressing numbers by words connoting, besides their normal meaning, objects, concepts, ideas or groups with a fixed number, technically known as *bhūta-saṅkhyā*, which was devised especially to facilitate referring to numbers in metrical compositions as well as for the sake of brevity and forms a unique feature of the Sanskritic language. This practice can be traced as early as the *Ṛgveda* (VII.103.1) and slightly later in the Brāhmaṇa and the Kalpa-sūtra texts.[31] They were employed not only to suit metrical exigencies but also to avoid repetition of the same word over and again as far as possible. Their use became especially popular from about the beginning of the Christian era when astrological-astronomical and mathematical treatises, involving repeated employment of numbers, came to be commonly composed in metres. Words actually connoting numbers (e.g. *eka, dvi, tri* etc.) as well as word-numerals or *bhūta-saṅkhyās* were often deployed together without any distinction. What is of special interest in the present context is that these word-numerals were employed in decimal place-value notation. It is pertinent to find in this context that the words signifying 'void', 'nothingness' or 'emptiness' were employed in this system for 'zero' and they included *kha, ākāśa, ambara, gagana, abhra, viyat, vyoma, antarikṣa, nabhas* (all denoting 'sky'), *śūnya* ('void'), *pūrṇa* ('full'), *randhra* ('hole') etc.[32]

However, zero is not always deployed directly and the necessity of its recurring use is obviated by the decimal place-value system. According to this system, the numbers are to be read and written from right to left (as against the normal Brāhmī *akṣaras* running from left to right) in keeping with the well-known principle *aṅkānāṃ vāmato gatiḥ* (numerals move [from right] to left). This principle invests every numeral sign on the left with ten times the value of that immediately to its right. Āryabhaṭa's mathematical system is based on the concept of decimal place-value which is defined by him as *sthānāt sthānaṁ daśa-guṇaṃ syāt*, viz. from one place to another (immediately to left) it is ten times.[33] Bhāskara's *Līlāvatī* calls it *daśa-guṇottara* (next being ten times [in value] as compared to the previous [right] one).[34] The date based on this sytem, as far as epigraphy is concerned, is met with for the first time in the Maṅkaṇi (Saṅkheḍā tālukā, Vaḍodarā district, Gujarat) plates of Taralasvāmin of the (Kalachuri) year 346 (594-95 or 595-96 A.D.) where the date is recorded following decimal place-value notation as 346 instead of the earlier method which would have required 300 (+) 40 (+) 6.[35] But even after the introduction of this system the inscriptional composers as well as writers/copyists of manuscripts continued to cling to the older method and the numbers are specified mostly according to the older system and only sparingly in the new decimal notation. It took a few centuries for the new numerical notation to finally replace the old notation. This reluctance to accept new developments is not typical of numbers alone and is noticed, *inter alia*, in the matter of change from one writing style to another as well.[36] Curious as it might look, we have instances where both the numeral and decimal place-value notations have been employed side by side in the same records in South Kosala (Chhattīsgaḍh region, Madhya Pradesh).[37]

Though in epigraphs the new decimal notation is known to have been employed first about the end of sixth century A.D., we have enough evidence of its much earlier use in astronomical-astrological literature. The celebrated astrologer-astronomer Varāhamihira (early sixth century A.D.) specifies dates and other numbers following this system in his

Pañcasiddhāntikā and *Bṛhat-saṃhitā*. Only a few examples would suffice. In the former text we have, *inter alia*,

(i) *sapta-aśvi-veda-saṅkhyā*[38] = 427 (Śaka)
(ii) *aṣṭa-yama-pakṣaiḥ*[39] = 228
(iii) *guṇa-kha-sapta*[40] = 703
(iv) *nava-vasu-guṇa-rasa-rasa*[41] = 66389
(v) *śara-nava-kha-indriy-ārṇava-aśāḥ*[42] = 1045095
(vi) *kha-indriya-diśaḥ*[43] = 1050

and in the latter one

(i) *nava-aṣṭa-pañca-aṣṭa*[44] = 8589
(ii) *śūnya-śarga-rāma*[45] = 3750
(iii) *ṣaḍ-dvika-pañca-dvi-yutaḥ*[46] = 2526.

Varāhamihira epitomises the contents of the five earlier Siddhānta texts, viz. Pauliśa, Romaka, Vaśiṣṭha, Saura and Paitāmaha and cites them besides the views of several earlier astronomers; and everywhere the numbers are indicated following the new decimal place-value notation. Unfortunately for us, none of these texts is available in its original form except only that of Āryabhaṭa (b. 476 A.D.) whose work specifies all numbers according to the decimal place-value system. While none of the original five Siddhāntas is now available, Bhaṭṭotpala in his commentary on the *Bṛhat-saṃhitā* mentions a text called *Puliśa-siddhānta* and cites at one place a stanza from what he terms *Mūla-Puliśa-siddhānta*.[47] All these indicate numbers following the decimal system.[48] In view of these facts it may be concluded that the new system based on decimal place-value principle had come into being a few centuries prior to the beginning of the Christian era though it took quite a few centuries to get popularised and many more to finally replace the old system.

It is noteworthy in the present context that all the three major elements contributing to the origin of the decimal place-value concept involving the placement of numbers from right to left (beginning with unit figure) and zero, were known from substantially earlier times and what was actually needed was only the use of zero in an arithmetical context and integrating the three into a system. As pointed out above, some scholars take the word *kṣudra* occurring in the *Atharva-veda* as denoting a negative or trifling number or zero. But as of now we have absolutely no evidence to show that it was actually employed in arithmetical calculations. In the various recensions of the *Yajurveda* we get the following numbers in a decimal order[49]: *eka* (= 1), *daśa* (= 10^1), *śata* (= 10^2), *sahasra* (= 10^3), *ayuta* (= 10^4), *niyuta* (= 10^5), *prayuta* (= 10^6), *arbuda* (= 10^7), *nyarbuda* (= 10^8), *samudra* (= 10^9), *madhya* (= 10^{10}), *anta* (= 10^{11}) and *parārdha* (= 10^{12}). The *Atharva-veda*[50] also specifies numbers up to 1000 in multiplication of 10.

Then, at least from the time of the Maurya emperor Aśoka (early third century B.C.) we have evidence of the practice of specifying higher numbers on left followed by lesser ones on right as in later times.[51]

The foregoing discussion would show that practically all the salient factors of the decimal place-value concept were known in their basic form from a hoary antiquity, but their conversion into a foolproof workable scheme and the consequent removal of the complexity of the older system took quite some time. It was certainly accomplished at least about a couple of centuries prior to the commencement of the Christian era with the employment of zero for an unspecified number and the application of the concept of *daśa-guṇ-ottara* which did away with the need for separate figures for tens (10 to 90), 100 and 1000. The entire arithemetical business could now be transacted with just ten figures, viz. 1 to 9 and 0.

It is felt in some quarters that the evolution of decimal place-value principle with zero as its hall-mark was actually not achieved independently by India and that it was in fact due to the Babylonians. O. Neugebauer, A. Sachs and lately David Pingree have pointed out that the system of numeration in the mathematical texts of the Old-Babylonian cuneiform records dating from about 1600 BC is based on place-value notation. Pingree has recently suggested that the Greeks borrowed this system from the Babylonians and 'Greek papyri of the period immediately preceding and following the beginning of the common (viz. Christian) era' show the use of a circular zero ($\bar{0}$ or δ) and that with the Greek influx in the early centuries of the Christian era these (place-value and zero) reached India.[52] However, despite early Indian contacts with Babylonia it cannot be argued that India had got these ideas from that quarter through Greek intermediaries. For, the Babylonians had sexagesimal notation while in India it was decimal place value-system. Moreover, there is clear literary evidence to show that Indians from at the latest second century BC were not only familiar with *śūnya* as a mathematical concept but had started using it in texts of other disciplines like metrics. Thus, what appears to be the fact is that both Indians and Babylonians developed the place-value notation independently of each other and probably Indians discovered *śūnya* somewhat earlier than the Greeks. That even the Arabs, spatially very close to the Babylonians, had to wait for several centuries to finally borrow Indian numbers would show that Indians were not aware of this Babylonian development.

The Arabs got familiarised with the Indian numerals first most probably in the latter half of the eighth century A.D. Works on Indian mathematics and astronomy including those of Brahmagupta (7th century A.D.) were rendered into Arabic during the enlightened reign of Caliph al-Manṣūr (755-75 A.D.) when Baghdād became a great literary and cultural centre. Thereafter numerous works dwelling upon Indian mathematics came to be written in Arabic. The most famous among such writers is Abū Jafar Muḥammad al-Khwārizmī who composed about 820 A.D. a work on Indian numerals which has come down to us only in its Latin translation called *Liber Algorisme de numero Indorum* which played a pivotal role in familiarizing the westerners with Indian numerical system.[53] The famous Arabic polymath, Abu Raihan ibn Muhammad, better known among historians as Al-Berūnī, in his *Tahquīkāt-e Hind* composed in 1030 A.D., admires the proficiency of the Hindus (Indians) in mathematics and observes "the numeral signs which we use are derived from the finest forms of the Hindu signs".[54] It took quite a few centuries for these Indian numbers to be finally accepted in Europe and it was only in the seventeenth century that they finally substituted the Roman letter-numerals, whereafter the system spread throughout the globe.[55]

REFERENCES

1. Up to 70000.
2. The earliest undoubted reference to numbers is met with in the records of the Maurya emperor Aśoka. It is possible that there are numeral signs in the Harappan script; but since despite all claims it remains still to be deciphered, it is difficult to be sure on this point. The Vedic and subsequent pre-Aśokan literature also contains references to numbers, sometimes exceptionally large.
3. Zero itself need not be employed and all the arithmetical numbers may be expressed by proper deployment of figures from 1 to 9 and zero, something impossible without implying the knowledge of zero.
4. This statement is true of tens, hundreds and thousands, there being no stroke or ligature appended to unit figures (1-9).
5. This is just a theoretical possibility. The highest such number encountered in

our records is 70000 occurring in a Nāsik inscription of the time of Nahapāna dating from first century AD. For Nahapāna's date, *vide* Ajay Mitra Shastri, "Śaka Era", *IJHS*, 31, pp. 75-77; *The Sātavāhanas and the Western Kshatrapas: A Historical Framework,* Nagpur, 1998, Part II, Chapter 2.

6. The symbols for 1 (–) and 2 (=) are generally appended to the top right portion of the figures for 100 and 1000 though there are quite a few instances where these signs are added about the middle or lower portion, and there are obviously no fixed rules on this point.
7. There must have been figures for higher numbers which are frequently mentioned in literary sources; but they don't appear in epigraphic and numismatic records. As stated above, the highest figure in a Brāhmī inscriptions is 70000.
8. We get three different forms of the figure for 200 in different versions of Aśoka's Minor Rock Edict I.

   Vedic literature including the Saṃhitā and Brāhmaṇa texts refers to extraordinarily large numbers which undoubtedly must have been written, though there is no direct reference to writing them. We have, however, explicit allusion to kine bearing the figure of '8' in their ears (*sahasraṃ me dadāto aṣṭakarṇyaḥ*) in the *Ṛgveda* (X.62. 7) being gifted to a priest by a ruling chief. There is left no doubt absolutely about this interpretation by Pāṇini (*Aṣṭādhyāyī*, VI.3.115) who also speaks of the number '8' together with a few other devices being marked on the ears of cows for identification.
9. *IA*, VI, p. 44. Earlier James Prinsep had conjectured that the Brāhmī numerals were the initial letters of the words denoting the numbers. This suggestion, favoured at one time by some European scholars, has, generally speaking, not much to commend itself and was never taken seriously by any scholar. This conjecture is applicable only to the numbers 1, 2 and 3 which have, *inter alia*, been indicated by letters *e, dvi* and *tri* respectively which are actually the initial *akṣaras* of the words connoting them. Likewise, *sva, sti* and *śrī* are employed in manuscripts for 1, 2 and 3 respectively which are actually the first, second and third letters of the auspicious formula *svastiśrī* often used at the beginning of any writing. Similarly, the letters *oṃ* (in a symbolic form), *na* and *maḥ* forming parts of the auspicious code *oṃ namaḥ* are employed for the three numbers in question.
10. *Ibid.*, VI, p. 148.
11. *Ibid.* Also his *Indian Palaeography*, Reprint, Calcutta, 1959, pp. 99-108.
12. *Bhāratīya Prāchīna Lipimālā* (Hindi), Ajmer, 1918, pp. 104-09. He has rightly pointed out that the signs for 1, 2, 3, 50, 80 and 90 do not resemble any of the letters and remained unaltered throughout.

13. *Vide* for a similar view, Shobhane Gokhale, *Indian Numerals*, Pune, 1966, pp. 47-48; K.V. Sarma, "Word and Alphabetic Numerical Systems in India", included in this volume.
14. For Āryabhaṭa's alphabetical system of numbers, *vide* his *Āryabhaṭīya*, Daśagītikā section, Paribhāṣā stanzas (I.2-3). Also see J.F. Fleet, "Āryabhaṭṭa's System of Expressing Numbers", *JRAS*, 1911, pp. 109-26; S.N. Sen, "Āryabhaṭa's Mathematics", *Bulletin of the National Institute of Sciences in India*, XXI (1963), pp. 298-302. But Āryabhaṭa's system did not attain popularity and none of the later writers is known to have followed it. This was, as pointed out by K.V. Sarma, due to the logically formed chronograms of this system 'not being amenable to be freely pronounced, copied with confidence, memorised with ease, suffixed with a case-ending and not fitted into a verse or formula'. *Vide* his paper mentioned in note 13 above. Āryabhaṭa appears to have been aware of an already existing system called *Kaṭapayādi* prevalent from at least the fourth century A.D. and perhaps devised his own system with the object of instilling greater brevity which, however, did not work. Later Āryabhaṭa II in his *Mahāsiddhānta* (*c.* 950 A.D.) made some improvement in the existing *Kaṭapayādi* system, the modified form of which is popularly called *Kaṭapayādi-II*. For the *Kaṭapayādi* system, *vide* J.F. Fleet, "The Kaṭapayādi System of Expressing Numbers", *JRAS*, 1911, pp. 788-94; K.V. Sarma, *op. cit.* For the use of *Dvādaśākṣarī* (*bārākhaḍī*) for expressing numbers prevalent in South India, Ceylon, Siam and Myanmar, see A.C. Burnell, *Elements of South Indian Palaeography*, Mangalore, 1878, p. 80; Bühler, *op. cit.*, pp. 106-7.
15. Burnell (*op. cit.*, p. 64) held that the serial order of the Brāhmī numerals had its origin in the Demotic system, while E.C. Bayley (*JRAS*, New Series, XIV, pp. 335ff.; XV, pp. 1ff.) thought that the Indian system was related in principle to the Hieroglyphic system even though the signs themselves were borrowed from the Phoenician, Bactrian and Akkadian figures or letters. According to Bühler, it is 'probable' that the Brāhmī 'numeral symbols were derived from the Egyptian Hieratic figures' (*op. cit.*, p. 101).
16. For a detailed criticism of these theories, *vide* Ojha, *op. cit.*, pp. 110-14; and for a comparative chart of the signs, *ibid.*, p. 113.
17. See S.N. Sen and A.K. Bag, "Vedic Mathematics", *The Cultural Heritage of India*, VI, eds. P.R. Ray and S.N. Sen, Calcutta, 1986, p. 34. *Vide* also Arun Kumar, *Devayāna: Saṃskṛti kā Vikāsa aura Veda* (Hindi), Faizabad, 1996, pp. 84-85, for a general account of Atharvavedic mathematics.
18. G. Coédes, "A propos de L'origine des chiffre arabes", *Bulletin of the School of Oriental and African Studies*, VI, pp. 323-28; L.C. Westenenkin, *Djava Periodical*,

I, p. 10; *Acta Orientalia,* II, pp. 12-21; H. Kern, "Inscriptie van Kota Kapur", *Bidragen tot de Taal-Land-en Volkenkunde van Netherlandische*, VII, pp. 208-9. These references are cited from *Cultural Heritage of India*, VI, p. 40, fn. 14; and A.K. Bag, "Symbol for Zero in Mathematical Notation in India", *Boletin de la Academia Nacional de Ciencias,* XLVIII, pp. 247-54, fns. 14-16.

19. For the origin and early History of the Śaka era, *vide* Ajay Mitra Shastri, "Śaka Era", *IJHS*, XXXI, pp. 67-88.
20. *Science and Civilisation in China*, III, Cambridge, 1959, p. 11. He also felt that 'the "emptiness" of Taoist mysticism, no less than the "void" of Indian philosophy, contributed to the invention of a symbol for *śūnya*'. *Vide ibid.*, p. 12.
21. Hira Lal, "Ragholi Plates of Jayavardhana II", *EI*, IX, p. 46, line 46 and end of the line 46 in the facsimile facing, p. 45.
22. V.V. Mirashi, "Kaman Stone Inscription", *ibid.*, XXIV, p. 336, lines 19 and 21 referring to numbers 220 and 180 respectively. In all the three cases the zero is represented by a dot.

    For numbers including zero met with in north Indian epigraphs, *vide* Om Prakash Lal Srivastava, *Uttara Bhārata mein Aṅkoṁ kā Vikāsa* (Hindi), Varanasi, 1986.
23. *dvir-ardhe/ rūpe śūnyam/ dviḥ śūnye/ tāvad= ardhe guṇitam/*

    Piṅgala-Chandaḥsūtra, ed. with Halāyudha's Mṛtiasañjīvanī commentary by Vishwanatha Shastri, Calcutta, 1874, VIII.29-31. Also *vide* Bibhutibhushan Datta and Awadhesh Narayan Singh, *History of Hindu Mathematics*, Mumbai, 1962, pp. 75ff. For a full discussion of the *Chandaḥsūtra* evidence on *śūnya, vide* S.R. Sarma, "*Śūnya*: Mathematical Aspect", *Kalātattvakośa: A Lexicon of Fundamental Concepts of Indian Arts*, Gen. Ed. Kapila Vatsyayan, Delhi, 1992, pp. 400ff.; "Śūnya in Piṅgala's Chandaḥsūtra", in this volume.
24. *viśvaṃ gaṇayato vidhātuḥ śaśi-kaṭhinī-khaṇḍena tamo-maṣī-śyāme-jina iva viyati saṃsārasy-ātiśūṅyatvāc= chūnya-bindava iva vilikhitāḥ . . . tārā arājanta.*

    Ed. and translation by Louis H. Gray, New York, 1930, pp. 99-100; S.N. Sen and A.K. Bag in *Cultural Heritage of India*, VI, p. 41 and fn. 18.
25. *Kalātattvakośa*, II, pp. 404-5, for the original citation and translation.
26. "Notices of Bakhshali Manuscript", *IA*, XII, pp. 89-90; "The Bakhshali Manuscript", *ibid.*, XVII, pp. 33-48.
27. B. Datta, "The Bakhshali Mathematics", *Bulletin of the Calcutta Mathematical Society*, XXI, pp. 1-60; S.N. Sen and A.K. Bag, *op. cit.*, p. 41; A.K. Bag, *op. cit.*, p. 250, however, places it in 400 A.D.
28. *Indian Palaeography*, p. 102.
29. It is not impossible that it may be a copy of an earlier work; but there is at present nothing to support this conjecture.

30. *Vide* S.N. Sen in *A Concise History of Sciences in India*, New Delhi, 1971, p. 178; A.K. Bag, *op. cit.*, pp. 250-51. However, the circular form is more common in these records.
31. For some instances of the Vedic usage of this practice, *vide* G.H. Ojha, *op. cit.*, p. 121; *The Cultural Heritage of India*, VI, p. 39.
32. For lists of a large number of word-numerals, *vide* G.H. Ojha, *op. cit.*, p. 120; G. Bühler, *op. cit.*, pp. 103-05. Over 800 such word-numerals have been listed by K.S. Shukla, *vide* his edition of *Vaṭeśvarasiddhānta*, Part I, New Delhi, 1986, Appendix. For a detailed discussion of this practice, *vide* K.V. Sarma, *op. cit.*, in this volume.
33. *Āryabhaṭīya*, Gaṇitapāda, verse 2.
34. *Daśa-guṇ-ottarāḥ saṃjñāḥ, Līlāvatī*, verses 10-11, for elaboration of this principle. In fact, the principle of *daśa-guṇ-ottara* is inconceivable without the knowledge of zero.
35. See V.V. Mirashi, *Inscriptions of the Kalachuri-Chedi Era, CII*, IV, Ootacamund, 1955, p. 164, line 20. The date is specified both in words and decimal figures: *saṃvatsara-śata-trayaṃ(ye) ṣaṭya(ca)tvāriṅ(ṃ)ś-ottarake 346.*

    Earlier only the second plate was known and its exact findspot uncertain. It was then edited by H.H. Dhruva under the title "A Gurjara Grant of Saṃvat 346" in *EI*, II, pp. 19-20. Since it was found in the Saṅkheḍā tālūkā, it was commonly called Saṅkheḍā plate. *Vide* Ojha, *op. cit.*, p. 115, fn. 1. The first missing plate of this charter was later reported from the village of Maṅkaṇi in the same tālukā and edited by A.S. Gadre in the *Important Inscriptions from the Baroda State*, I, pp. 4f. V.V. Mirashi republished the entire record in his aforesaid work (pp. 160-65) under the caption "Spurious Mankani Plates of Taralasvāmin: (Kalachuri) Year 346". As will be clear from the title, he regards the record as 'spurious' and the reasons leading to this view are discussed by him in a paper published in the *Journal of the Gaṅganatha Jha Research Institute*, II, pp. 387 ff and are found reiterated in *CII*, IV, pp. 161-63. In his opinion, it was 'forged some time in the second half of the eighth century A.D. when the decimal notation came into vogue' (*ibid.*, p. 162). However, the arguments of Mirashi in support of his contention are not at all convincing. We have discussed this problem at length in our paper under publication in the *Annals of the Bhandarkar Oriental Research Institute.*
36. E.g. the box-headed characters had gone out of vogue and the current form of Brāhmī in use in South Kośala was nail-headed variety with southern characteristics as seen from the inscriptions of the period of the Pāṇḍuvaṃśins engraved on stone. However, their official records on copper plates continued

to be engraved in the box-headed variety due to long-established scribal tradition which the royal scribes found difficult to abandon while others had already adopted the new writing-styles as they wanted to be in tune with the changed times and to make their writing easily readable to the literate people.

37. The Mallār plates of the Amarārya-kula king Vyāghrarāja (Ajay Mitra Shastri, *Inscriptions of the Śarabhapurīyas, Pāṇḍuvaṁśins and Somavaṁśins*, Part II, New Delhi-Delhi, 1995, Pl. XLIII, third plate, line 4) give the year according to the decimal principle and *tithi* following the older system. For discussion, *vide ibid.*, pp. 63-64, fn. 7. Likewise, the Lodhiā plates of the Pāṇḍuvaṃśin king Śivagupta Bālārjuna, dated year 57, specify the year according to the old practice and the *tithi* (30) contains a dot after 30 (*vide ibid.*, p. 131, line 32, Plate LXXV, third plate). Thus both the systems are found to be adopted in the same records which represent the transitional phase.

Earlier, the Mankuwar image inscription of the Gupta emperor Kumāra Gupta I's reign dated in Gupta year 129 (448-49 A.D.) shows a confusion between the older system and the new one by drawing the figure for 100 in a strange manner and including a figure for zero connected by a horizontal stroke to left in the middle. For the inscription, *vide* J.F. Fleet, *Inscriptions of the Early Gupta Kings and their Successors, CII*, III, Calcutta, 1888, pp. 46-47; and for a discussion of the formation of the date, *vide* B.N. Mukherjee, "The Early Use of Decimal Notation in Indian Epigraphs", *JESI*, XIX, pp. 80-81.

As shown by Subrata Kumar Acharya ("The Transition from the Numerical Form to Decimal System in the Inscriptions of Orissa", *ibid.*, XIX, pp. 52-62), in Orissa the transition was effected by the Eastern Gaṅgas from about mid-seventh century A.D. and perfected by about the close of the same century. In Bangladesh the Ashrafpur inscription of the Khaḍga king Devakhaḍga of the year 73 of the Harṣa era (= 679 A.D.) is the earliest epigraph showing the employment of the decimal notation even though the *tithi* is cited according to the old system. *Vide* B.N. Mukherjee, *op. cit.*, pp. 80-84.

What deserves special notice in this connection is the occurrence of dates based on decimal notation as early as the seventh-eighth centuries A.D. in Sanskrit inscriptions from Jāvā (Dinaya), Bakul and Po-nagar (Champā) in South-East Asia (*supra*) which would indicate that the areas from where the Indians had migrated to these regions had adopted the decimal place-value notation completely and that the Indian immigrants adhered to this system religiously.

38. *Pañca-siddhāntikā*, with an English translation and notes by T.S'Kuppanna Sastry and edited with an introductiion and appendices by K.V. Sarma, Chennai, 1993, Ch. I, verse 8.

39. *Ibid.*, I.9.
40. *Ibid.*, I. 10.
41. *Ibid.*, I.14.
42. *Ibid.*
43. *Ibid.*, I. 15.
44. *Bṛhat-saṃhitā*, with Bhaṭṭotpala's Vivṛti commentary, edited by Sudhakara Dvivedi, Vijayanagaram Sanskrit Series, Varanasi, 1895, VIII.20.
45. *Ibid.*
46. *Ibid.*, XIII.3.
47. *Vide* for these citations, Shankar Balakrishna Dikshit, *Bhāratīya Jyotiṣaśāstra* (Marathi), 2nd edn., Pune, 1931, pp. 162 and 163; G.H. Ojha, *op. cit.*, p. 116, fn. 7.
48. Both with zero and with zero implied and often fairly large numbers.
49. *Yajurveda*, XVII.2.
50. V.15. 1-2.
51. As suggested by the expressions like *ekādaśa* (11), *ekaviṃśati* (21) etc., this practice and the principle of writing numbers from right to left in ascending order must have been in vogue right from Vedic times when we start getting them.
52. O. Neugebauer, *The Exact Sciences in Antiquity*, Copenhagen, 19511, p. 20. For a summary of these views, see David Pingree, "Zero and the Symbol for Zero in Early Sexagesimal and Decimal Place-Value Systems", included in this volume.
53. For a detailed summary of various views on this phenomenon, see R.C. Gupta, "Spread and Triumph of Indian Numerals", *IJHS*, XVIII, pp. 23-38.
54. E. Sachau, *Al-Beruni's India*, I, London, 1910, p. 174. Also p. 177 for some other details.
55. For this, *vide* L.C. Gupta, *op. cit.*

ABBREVIATIONS

CII	*Corpus Inscriptionum Indicarum*
EI	*Epigraphia Indica*
IA	*Indian Antiquary*
IJHS	*Indian Journal of History of Science*
JESI	*Journal of Epigraphical Society of India*
JRAS	*Journal of the Royal Asiatic Society*

Kharoshṭī Numerals and the Early Use of Decimal Notation in Indian Epigraphs

B.N. MUKHERJEE

A

The Kharoshṭī alphabet was probably created during the period of the early Achaemenid rule in the north-western section of the Indian subcontinent, perhaps in the late 6th century B.C. or early 5th century B.C. (though the earliest known long Kharoshṭī inscriptions should be dated to the 3rd century B.C.).[1] Out of twenty-two Aramaic characters as many as twelve (or sixteen) were adopted (in some cases with modifications) in the new script. Similarly, modified forms of five (or eight) Brāhmī letters were accepted. This evidence should incidentally suggest the existence of Brāhmī in c. 6th century B.C. The rest of the characters and signs for medial vowels, super and superscript *r*, subscript *y* etc., were ingeniously devised (perhaps under influence of forms of some Aramaic and Brāhmī letters and relevant signs in Brāhmī). Again, in case of vowels, medial signs were added to a basic vowel letter (*a*) to express initial vowel characters.[2] Like Aramaic, Kharoshṭī was written from the right to left.[3] But the number of letters were more numerous in Kharoshṭī than in Aramaic.[4]

It appears that the alphabet was devised thoughtfully to help the Aramaic knowing officials of the Achaemenid empire to write North-Western Prākṛt.[5] Later, by the time of the Maurya emperor Aśoka (c. 273-236 B.C.), it had become a popular script.[6]

Like the letters, the Kharoshṭī numerals had primarily a Semitic origin. As in Aramaic, the numerals 1 to 5 (and perhaps up to 9) were marked by vertical strokes.[7] However, later, perhaps in the Scytho-

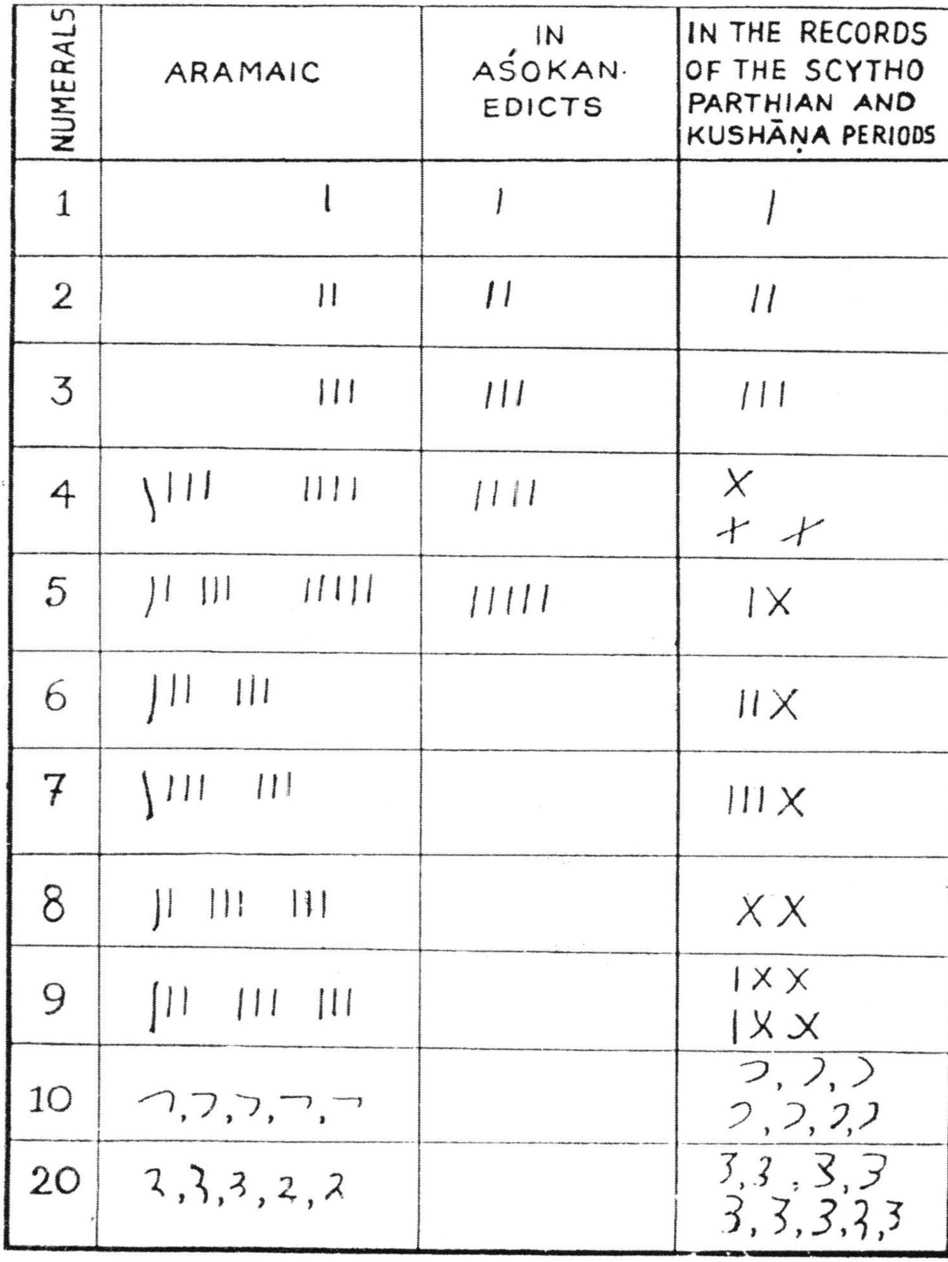

Fig. 1. Kharoshṭī Numerals (Indian Subcontinent)

Parthian age, the numeral 4 was expressed by a slanting cross (X).[8] Here we may discern an influence of Brāhmī (see the straight cross-like (+) form of this numeral in Aśoka's thirteenth rock edict at Kalsi).[9] The forms of the numerals 10 and 20 betray influence of their forms in Aramaic.[10] Unlike in early Brāhmī, there were no signs to stand for 30, 40, 50, 60, 70, 80 or 90. For expressing any figure up to 99, one had to depend on one or a combination of the above noted numerals (Fig. 1 Nos. 1-20).

None of the attested forms of the symbol of figure for 100 in Kharoshṭī can be placed before the 1st century A.D. (or very late 1st century B.C.). But during the earlier centuries of the existence of the script concerned (prior to the beginning of the Christian era), the form (or forms) of the figure for hundred could have undergone changes, as were in the cases of the appearances of certains letters (like *da, ya, śa, sa* etc.). So in the 1st century A.D. somewhat developed forms of the figure for 100 were expected to be witnessed in the records concerned. Nevertheless, slight differences between different developed forms may help us in tracing the line of development in this regard (Figs. 2 & 3 Nos. 1-16).

A comparative study of these forms convinces one that the sign for hundred in Kharoshṭī, as used in the 1st century A.D. or later, could not have possibly originated from the shape (or shapes?) of the symbol for hundred as appearing in the Aramaic edicts of Aśoka (c. 273-236 B.C.) ().[11] While the latter can be traced from one of the forms of the figure for hundred in the Aramaic version of Behistan (or Bisitun) inscription of the Achaemenid emperor Darius I (c. 522-486 B.C.) (),[12] the form(s) of the figure for 100 in Aśokan Aramaic could not have constituted an intermediary stage in the progress of the formation of the symbol for hundred in Kharoshṭī. On the other hand, there is a separate form of the figure for 100 in the above noted Aramaic inscriptions of Darius which may be considered in this regard. From the readers' point of view, the symbol in question looks like an upturned "U"-shaped figure (with its right vertical slightly shorter than the left vertical stroke) having a slanting stroke (with a north-east to south-west orientation) attached to the left side of its topmost (apsidal-like) section, ().[13] The line of formulation of the shape of Kharoshṭī 100 can be traced by a comparison of this figure with the slight differences between its developed forms in the early centuries A.D. (Figs. 2-3).

A close scrutiny of the attested forms of the figure for Kharoshṭī 100

I. Forms of the figure for 100 in Aramaic (c. 6th. Century B.C. to c. 3rd. Century B.C.)	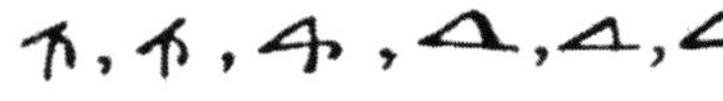 etc.
II. Early forms of the Brāhmī letter ta	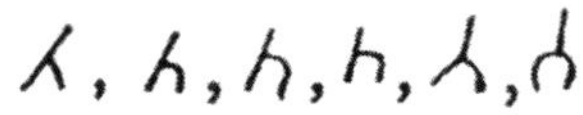 etc.
III. Some early forms of the Aramaic letter tāw (t, ṯ) (c. 7 th. century B.C. to c. 3 rd. century B.C.)	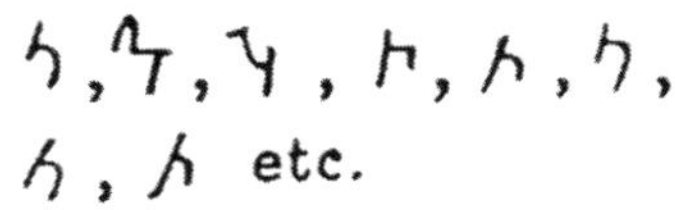 etc.

Nº	INSCRIPTION	DATE	FORM
1	Peshawar	Year 168 = c. 2 B.C. +	
2	Khalatse	Year 187 = c. A.D. 17 +	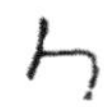
3	Sirkap	Year 191 = c. A.D. 21 +	
4	Dewai	Year 200 = c. A.D. 30 +	
5	Dasht-e-Nawur	Year 279 = c. A.D. 32 ×	
6	Mount Banj	Year 102 = c. A.D. 43–44 ⊕	
7	Takht-i-Bahi	Year 103 = c. A.D. 44–45 ⊕	
8	Paja	Year 111 = c. A.D. 52–53 ⊕	

Fig. 2. The forms of the figure for 100 in Kharoshṭī Inscriptions.
(Indian Subcontinent)

Nº	INSCRIPTION	DATE	FORM
	CONTD.		
9	Kaldarra	Year 113 = c.A.D. 54–55 ⊕	
10	Panjtar	Year 122 = c.A.D. 63–64 ⊕	
11	Kalawan	Year 134 = c.A.D. 75–76 ⊕	
12	Taxila	Year 136 = c.A.D. 77–78 ⊕	
13	Kula Dheri (Charsadaa)	Year 303 = c.A.D. 133 +	
14	Lorian Tangai	Year 318 = c.A.D: 148 +	
15	Jamalgarhi	Year 359 = c.A.D. 189 +	
16	Hastnagar	Year 384 = c.A.D. 214 +	

INDEX:

\+ *Dated in the Era of c. 170 B.C.*

× *Dated in the Parthian Era of 247 B.C.*

⊕ *Dated in the Era of 58 B.C.*

Fig. 3. The forms of the figure for 100 in Kharoshṭī Inscriptions. (Indian Subcontinent)

indicates that gradually the left-hand curved line was shortened and abolished. On the other hand, the right-hand curved line was lengthened vertically (sometimes with a slant to the left) (as in the case of the Peshawar inscription of the year 168, dated in the Old Śaka era of c. 170 B.C., or in c. 2 B.C. ( ).[14] In certain inscriptions the horizontal section of the right-hand stroke was lengthened and its vertical elongation was kept short (as in the Khalatse inscription of the year 187 of the Old Śaka era of 170 B.C., or of c. A.D. 17)[15] ( ). In a few cases a slanting stroke was added to the bottom of the curved line (as in the Sirkap inscription of the year 191 of the Old Śaka era, or of c. A.D. 21)[16] ( ). In some records the right hand curved line was transformed into a slanting line (with an inclination from the north-west to south-eastward) (as to be seen in the Takht-i-Bahi record of the year 103, or of c. A.D. 44-45)[17] ( ). In this form the figure has striking resemblance to an early form of Brāhmī *ta* ( ).[18] Occasionally the right-hand line was continued upwards through a small curved stroke across the slanting line (as in the Taxila inscription of the year 136) ( ).[19] However, the general tendency was, as pointed out above, to elongate the curved line. Gradually it was straightened and the upper stroke was slightly elongated and often slightly curved at the lower end and/or dented in middle (, , etc.).[20] In this form the figure appeared like the letter *'e* in Kharoshṭī (i.e. *'a* with the sign for medial *e* ( ).[21]

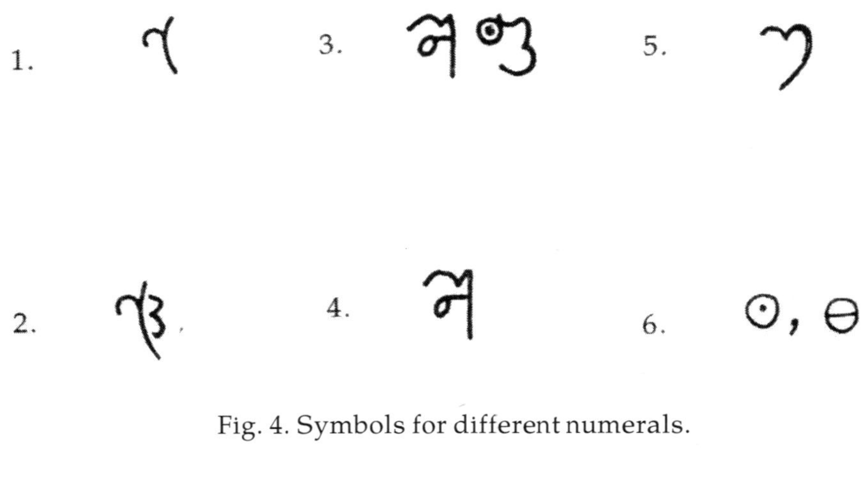

Fig. 4. Symbols for different numerals.

Besides this similarity certain forms of Kharoshṭī 100 had at least superficial resemblance to Aramaic *tāw* (*t*) (𐡕, 𐡕),[22] or, as noted above, with an early form of Brāhmī *ta* (𑀢).[23] But, since the development of the figure of Kharoshṭī 100 can be logically shown to have been derived from a form of the same number in Aramaic of the Achaemenid empire in the late 6th century B.C.,[24] the connection between the two should be accepted as historically correct. This inference strongly supports the hypothesis that Kharoshṭī originated in the early Achaemenid age (late 6th–early 5th century B.C.).[25]

The value for each hundred was indicated by putting the figure or figures of the necessary numeral or numerals on the right of the sign for hundred (from the readers' point of view). The numerals indicating figures below one hundred were used to be written on the left of the figure for hundred. Thus the figures denoting "297" would have been expressed as III X 73333Ƭ II. In this system of writing numerals we may also discern Semitic influence.[26]

The dates in Kharoshṭī documents never indicate any sense of decimal notation. There was never a sign for zero (*śūnya*) in Kharoshṭī. Morover, the fact that Kharoshṭī of the North-West was in regular use up to *c*. 4th century A.D. in one of the most cultured zones of the Indian subcontinent, should suggest that the popular use of the sign for zero had not developed in India till that period.

B

On the other hand, the use of decimal notations in expressing dates in Brāhmī (or in one of its derivative scripts), became a regular feature in epigraphs from about the 9th century A.D. in the Indian subcontinent or at least in various parts of it.[27] However, expressions of dates in decimal figures can be noticed occasionally from the 7th century A.D. There are

Fig. 5. The numerical symbols for expressing the year "129" in the Mankuwar inscription.

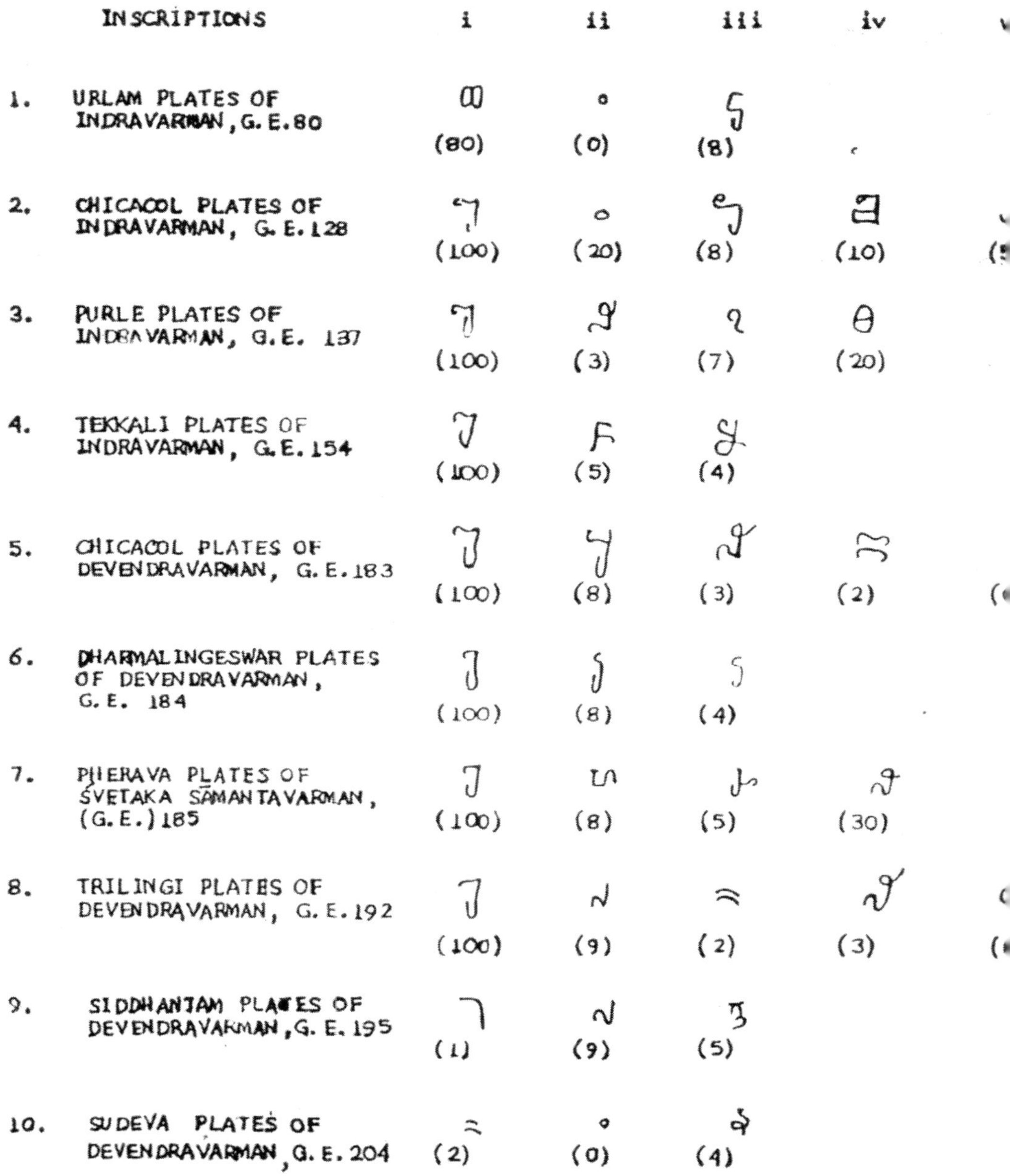

Fig. 6. Indications of the use of the decimal system of notation in certain inscriptions of the Eastern Gaṅgas (The drawings of figures have been adopted from an article published in the *Journal of the Epigraphical Society of India,* vol. XIX, p. 61).

also well-known instances of the use of decimal notation (expressing "zero" by a dot or a circle) in a few South-East Asian inscriptions of the early 7th century of the Śaka era (Śaka years 605, 606 and 608).[28] These instances allude to the import of the system from India, where the Śaka era as well as the script used in the epigraphs concerned originated.[29]

In India an attempt to introduce the system can be noticed in an inscription on the pedestal of an image of the Buddha, found at Mankuwar (Allahabad district, U.P.).[30] It is dated in the Gupta year 129 or A.D. 448-49.[31] The date is expressed by numerical symbols,[32] to be deciphered as 100 (+) 20 (+) 9.[33] But the symbol for the year "100" does not have here one of its well-known contemporary forms (Fig. 4, no. 5).[34] It is drawn in a manner which suggests the inclusion within it the figure for "20" or "zero" (0) (Fig. 4, no. 4). Again, the symbol for "9" is seen connected by a curved line with that for "20" (Fig. 4, nos. 3 and 4).[35] These departures from the norms may perhaps betray the engraver's or the original writer's attempts to transform the knowledge of decimal notation into practice (Fig. 5).

However, the regular use of the decimal system (with the help of the figures for 1 to 9 and zero) was not known to Āryabhaṭa (I) (b. A.D. 476), the famous mathematician and astronomer and the author of the *Āryabhāṭīya* (A.D. 499).[36] Hence the system in question did not probably gain ground till the late 5th century A.D. This inference holds good even

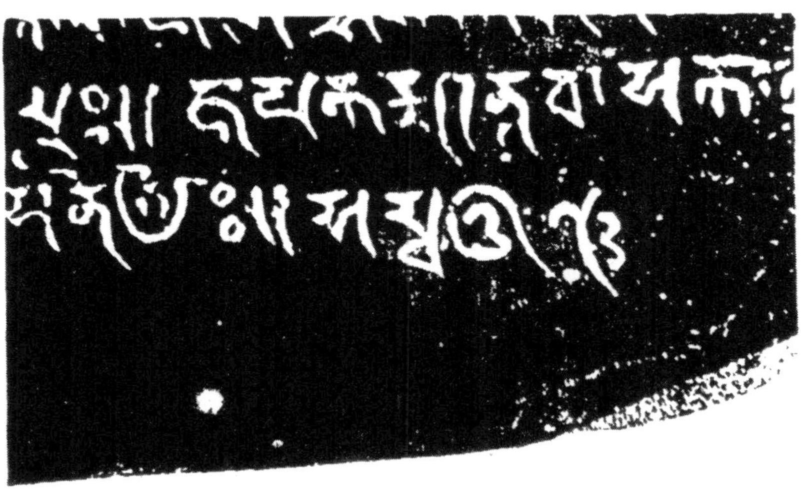

Fig. 7. The year "73" in a inscription of the 7th century found at Asrafpur.

after conceding that the concept of zero might have been known in India from a much earlier age.[37]

In several inscriptions of the Eastern Gaṅgas, the symbol for "100" is followed by the figures betraying the attempts to use the decimal system of notation.[38] Thus in the Chicacole plates of Devendravarman the year expressed as 100 (+) 83 really stands for 183.[39] The inclination for employing this system is suggested by the figure of zero inserted between the symbols for 80 and 3 in another record.[40] Obviously here the intended number is "83" (Fig. 6).

The correct use of cipher is indicated by the Srikakulam plates of the Gaṅga Year 187, in which "20 has been written with 2 and 0 in accordance with the decimal practice".[41] In the Siddhantam plates of Devendravarman the Gaṅga year 195 is expressed by the figures for 1,9 and 5 (Fig. 6).[42] Since the Gaṅga era commenced sometime in the period from 496 and 498,[43] the decimal system must have been used in dating in parts of India by the second half of the 7th century A.D. The same indication is given by the year 73 (of the Harṣa era of A.D. 606), expressed by the figures 7 and 3 in a copper plate inscription of Devakhaḍga found long ago at Asrafpur in then Dacca district (now in Bangladesh)[44] (Fig. 4, nos. 1 and 2; Fig. 7).

A still earlier use of the decimal notation can be found in the Mankani inscription of Taralasvāmin of the year 346, assignable to the era of A.D. 249 and so datable to c. A.D. 595.[45] Here the date is expressed by the figures 3, 4 and 6. Unfortunately, this epigraph is generally considered as a spurious record.[46]

Thus, in present state of our knowledge, we may hold that the use of the decimal notation was practised in at least certain areas of India by about the second half of the 7th century A.D. As noted above, its use in Indian epigraphs became a regular feature in several parts of the subcontinent from about the 9th century A.D.[47] On the other hand, the first attempts at introducing this system might have occurred in the 5th century A.D. The decimal notation was known to Varāhamihira (early 6th century A.D.).[48] The *Vāsavadattā* of Subandhu, which was well-known by the early 7th century A.D., refers to "zero" as *śūnyabindu* ("the void expressed by a dot").[49] The emergence of the concept of "zero" itself preceded, as indicated above, all these dates by a long period of time.[50]

REFERENCES

1. The correct spelling of the script concerned is *Kharostī* or *Kharoshṭī*. The latter form occurs in the *Lalitavistara* (10th *adhyāya*; P.L. Vaidya editor , *Lalitavistara*, Darbhanga, 1958, p. 88). The name was wrongly read as Kharoshṭhī by G. Bühler in the last quarter of the 19th century. This misreading was repeated by scholars until it was pointed out in 1980 (*Monthly Bulletin of The Asiatic Society*, May, 1980, p. 4; *Journal of the Asiatic Society*, 1981, pp. 144f). Unfortunately some scholars still persist in repeating the mistake.
   The only suitable period for the creation of the Kharoshṭī script was the time of the dominance of the Indo-Iranian borderlands by the Achaemenids, in whose empire the Aramaic script and the Aramaic language had an official status (G. Bühler, *Indian Palaeography* [English translation], reprint, Calcutta, 1959, p. 37; A.H. Dani, *Indian Palaeography*, Oxford, 1963, p. 254).
2. Either Aramaic or Brāhmī influence may be claimed to have worked behind the creation of a number of Kharoshṭī characters (like *ga, ṇa* and *śa*). The Aramaic impact on the form of Kharoshṭī' *a* is possible, but not certain (see one of the forms of Aramaic *'aleph* used in the Kandhar and Lamghan I edicts of Aśoka) (B.N. Mukherjee, *Studies in the Aramaic Edicts of Aśoka*, Calcutta, 1984, [cited below as *SAEA*], pl. x, cols. 3 and 10, row. 1).
3. A.H. Dani, *op. cit.*, p. 235.
4. The creation of new letters and conjuncts was necessary for writing Prākṛt. The number of characters increased in Central Asian Kharoshṭī and to a certain extent, in Eastern Kharoshṭī (B.N. Mukherjee, *India in Early Central Asia*, New Delhi, 1995, pls. 4 and 5; *Kharoshṭī and Kharoshṭī-Brāhmī Inscriptions in West Bengal* [*India*], Calcutta, 1990, pls. IIf).
5. A.H. Dani, *op. cit.*, pp. 255-56.
6. Unless the Kharoshṭī script was understood by a substantial number of the Prākṛt speaking people in the north-west of the Indian subcontinent, Aśoka would not have used it for writing his Prākṛt edicts in that area. His edicts were meant to be read (or heard) by the subjects of his empire.
7. H. Hultzsch, *Corpus Inscriptionum Indicarum, Inscriptions of Aśoka*, Oxford, 1925, pls. facing 56, 68 and 78; G. Bühler, *op. cit.*, *Chart*, pl. I; Col. XIII, rows 1-5. In the Teima inscription, the figure for 10 showed a form close to the appearance of the same numeral in Kharoshṭī. The sign for "20" in Kharoshṭī resembled the form of the same numeral in some 6th and 5th centuries B.C. documents. In early Aramaic, the numerals "1 to 10" (or 1 to 9?) were marked by vertical strokes. In the 3rd century B.C., a horizontal line slightly curved at one end

denoted "10" in Aramaic (G. Bühler, *op. cit.*, p. 96). In this connection see also *SAEA*, figs. 1, 7 and 10; see also below n. 13.

8. S. Know, *Corpus Inscriptionum Indicarum*, vol. II, pt. I, *Kharoshṭī Inscriptions with the Exception of those of Aśoka*, Calcutta, 1929 (cited below as CII, vol. II, pt.I), pl. V, no. 1, 1.1.
9. H. Hultzsch, *op. cit.*, pl. 50, 1.6.
10. G. Bühler, *op. cit.*, p. 96; *SAEA*, pl. X, cols. 9 and 10, rows 24 and 25. See also above n. 7-8.
11. *SAEA*, fig. 1, *l*.4. The form is clear in the first Lamghan edict, but not in the second Lamghan edict (fig. 2).
12. J. Greenfield and B. Porten, *The Bisitun Inscription of the Darius the Great—Aramaic Version*, London, 1982, pl. III, 1.11.
13. *Ibid.*, pl. III, 1.17.
14. *CII*, vol. II, pt. 1, pl. XV, no. 1; B.N. Mukherjee, *Central and South Asian Documents of the Old Śaka Era*, Varanasi, 1973 (cited below as *CSADOSE*, p. 98).
15. *CII*, vol. II, pt. 1, pl. XV, no. 2; *East and West*, vol. IX, no. 4, 1958, p. 294, fig. 8; *CSADOSE*, p. 98. The figure for 100 is more clearly discernible in the reproduction of the inscription in the *East and West* than in *CII*, vol. II, pt. I.
16. *CII*, vol. II, pt. 1, pl. XVI, no. d; *CSADOSE*, p. 98.
17. *CII*, vol. II, pt. 1, pl. XII, no. 1; *CSADOSE*, p. 108.
18. A.H. Dani, *op. cit.*, pl. IIIa, row 1.
19. *Epigraphia Indica*, vol. XXI, pl. facing p. 259; *CII*, vol. II, pt. 1, pl. XIV.
20. *CII*, vol. II, pt. 1, pl. XXII, 1 and 10; pl. XIII, no. 4 etc.
21. A.H. Dani, *op. cit.*, pl.XXIIIa, row 1.
22. D. Diringer, *The Alphabet,* London, reprint, 1953, p. 258, fig. 127.
23. See above n. 18.
24. The major part of the Behistan inscription of Darius I is dated to period from 520 and 518 B.C.
25. See above n. 1.
26. G. Bühler, *op. cit.*, pp. 96-97. The Kharoshṭī practice of prefixing the number of century or centuries to the symbol for 100 "is found in all Semitic systems of numeral notations" (G. Bühler, *op. cit.*, p. 97). In Central Asian Kharoshṭī the figure for 1000 was expressed by two forms (*India in Early Central Asia,* pl. 5 [b]).
27. G. Bühler, *op. cit.*, p. 102.
28. *Bulletin of the School of Oriental Studies*, vol. VI, pp. 323f; D.M. Bose, S.N. Sen and B.V. Subbarayappa (editors), *A Concise History of Science in India*, Delhi, 1971, p. 178.

29. *Ibid.* This evidence suggests the Indian and not the Chinese inspiration behind the introduction of decimal notation in S.E. Asia.
30. D.R. Bhandarkar (revisor), *Corpus Inscriptionum Indicarum*, vol. III, *Inscriptions of the Early Gupta Kings*, New Delhi, 1981, p. 291.
31. *Ibid.*, pp. 292-93.
32. *Ibid.*, pl. XXV.
33. *Ibid.*, p. 293.
34. S. Gokhale, *Indian Numerals,* Poona, 1906, p. 19.
35. D.R. Bhandarkar, *op. cit.*, pl. XXV. For another view see the *Journal of Ancient Indian History,* vol. IV, pts. 1-2, p. 127.
36. M. Winternitz, *History of Indian Literature*, vol. III, pt. II (translated by S. Jha), Delhi, 1967, p. 649; D.M. Bose, S.N. Sen and B.V. Subarayappa (editors), *op. cit.*, pp. 51; *Journal of Ancient Indian History*, vol. IV, pts. 1-2, p. 100. It should, however, be noted that Āryabhaṭa's alphabetical system of expressing numbers with the help of consonants and vowels was based "on the decimal place-value principle" (*Āryabhāṭīya, Daśagītikā*, if; D.M. Bose and others, *op. cit.*, p. 174).
37. For an example we may refer to the *Chandaḥ-sūtra* of Piṅgala (see also D.M. Bose and others, *op. cit.*, p. 175).
38. *Journal of the Epigraphical Society of India* (cited below as *JESI*).
39. *Ibid.*, p. 61.
40. *Ibid.*
41. *Ibid.; Journal of Ancient Indian History* vol. IV, pts. 1-2, p. 127.
42. *JESI*, vol. XIX, p. 61.
43. D.C. Sircar, *Indian Epigraphy*, Delhi, 1965, p. 291.
44. *JESI*, vol. XIX, p. 80. The date was earlier wrongly read as 10 (+)3 (*ibid.*).
45. N.V. Mirashi, *Corpus Inscriptionum Indicarum*, vol. IV, pt. I, Ootacamund, 1955, pp. 161-64; *JESI*, vol. XIX, p. 80.
46. *Ibid.*
47. *JESI*, vol. XIX, p. 80.
48. *Pañcasiddhāntikā*, 1, 8; 8.1; 18, 33; M. Winternitz, *op. cit.*, p. 646; R.C. Majumdar (editor), *Classical Age,* Bombay, 1954, p. 321.
49. R.C. Majumdar, *Classical Age*, p. 316; *Vāsavadattā* (edited by F.E. Hall), p. 182.
50. See above, n. 37.

    The Bakhshali manuscript, which uses decimal notation, should be palaeographically dated to c. 8th-9th century A.D. (*Indian Antiquary,* vol. XVII, p. 36; D.B. Pandey, *The Shāhis of Afghanistan and the Punjab*, Delhi, 1973, pp. 137f) and not to any earlier age. Nevertheless, its original text could have been produced in an earlier age. On the other hand, there were possibilities of updating the text. So the evidence of the manuscript concerned cannot be used by us.

The dates of some inscriptions from Gujarat, noted in the 4th or early 5th century of the Gupta era, are recorded in decimal figures in the printed texts or notices (*Journal of the Asiatic Society of Bengal*, vol. VII, p. 968; *Progress Report of the Archaeological Survey of India, Western Circle*, 1915-16, p. 55; D.R. Bhandarkar, *A List of Inscriptions of Northern India*, ncs. 1357 and 1361). But these cannot be verified for want of facsimiles. Moreover, as V.V. Mirashi has pointed out, in several inscriptions from the same area and dated in the later years of the Gupta era, the dates are expressed in numerical symbols and not in decimal figures (V.V. Mirashi, *op.cit.*, pt. I., p. 162, f.n. 3.

Reflections on Pāṇinian Zero*

M.D. PANDIT

INTRODUCTORY

Pāṇini's grammar is a masterpiece of the technical skill combined with almost total comprehension of the data on which he is working and has stood the test of time for all these 2,500 years. As such, it has been a subject of grammatical studies in India and abroad. It has exercised such an impact of perfection and authority that grammatical studies came to be identified with the study of only Pāṇini's *Aṣṭādhyāyī* in ancient and even in modern India. The western scholars who first studied Pāṇini were impressed by his technical skill and scientific accuracy and perfection.

Pāṇini has used many techniques like *it-saṃjñā, ādeśa, āgama,* technical terms like *ghi, ghu, ti* etc. which lock apparently meaningless. From the usage of such a highly developed technique in Pāṇini's grammar, one is always tempted by the idea of a comparison of the Pāṇinian techniques with those adopted in sciences like modern mathematics, computer or even symbolic logic. The technique of *lopa* or zero which Pāṇini has employed in his grammar for descriptive purposes is one such technique.

The scope of the present paper is to study this technique of zero used by Pāṇini in its entirety. As we go through Pāṇini's grammar we find a zero employed in the case of the form *atti*, from √*ad,* 2nd conj. 3rd *pers.* sing. We also find that there are some sounds used by the Śiva–sūtras and by Pāṇini which are termed as the *it*-sounds. Though their

* For a detailed discussion and treatment of the subject, cf. M.D. Pandit, *Zero in Pāṇini*, Centre of Advanced Study in Sanskrit, University of Poona, Pune, 1990.

presence is visible on the arbitrary, technical plane, they are made to go away afterwards. Pāṇini has also used the technique of *anuvṛtti* in which some common words from the preceding *sūtras* are dropped out from the succeeding sūtras. Looked from the point of view of the technique of axiomatics, the words which are dropped out are nothing but zeros. So we have three types of zero-technique used by Pāṇini in his *Aṣṭādhyāyī*. The present work aims at the study of all these three types of zeros from different angles and attempts to find out the possible origin of this technique.

TWO MAIN LEVELS IN PĀṆINI'S GRAMMAR

In all we have two main levels on which the technique of 'zero' is working in Pāṇini's grammar: (1) the level of linguistic description or the linguistic level; (2) and purely technical level. In the case of technical level also, we have two divisions — (a) the technique of *it-saṃjñā* and (b) the technique of *anuvṛtti*. The following figure will give the correct picture of these divisions at a glance :—

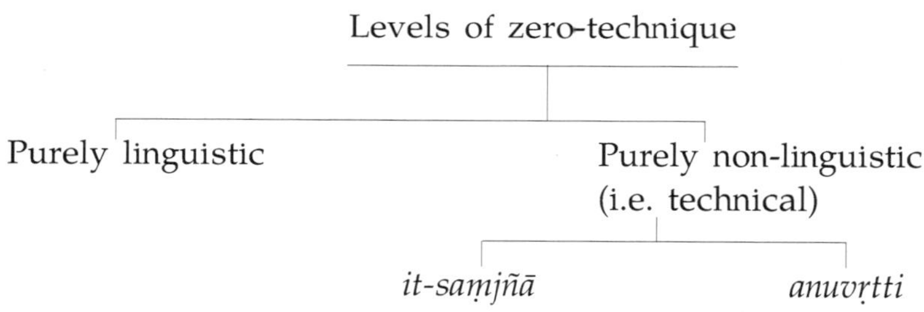

The following examples from Pāṇini will illustrate the point.

EXAMPLE FROM THE LEVEL OF LINGUISTIC DESCRIPTION

Let us take an example from Sanskrit, say, of roots √*khād* and √*ad* (2nd conj.) both meaning 'to eat'. The grammatical statement of pres. 3rd pers. sing of √*khād* according to Pāṇini is as follows :

khād + śap + ti
=khād + ta + ti
=khādati.

In the case of √*ad*, which being in the 2nd conjugation takes no *vikaraṇa* or middle morpheme, it is as follows :

ad + ti
= *ad + śap + ti*
= *at + ti*
= *atti,* the *vikaraṇa śap* being lost according to *adiprabhṛtibhyaḥ śapaḥ,* 2.4.72.

Here it is quite clear from the analysis of the two forms that while the form *khādati* has one *a* between the root √*khād* and the termination *ti,* the form *atti* has no such infix. There are, therefore, two ways for accounting for the forms: (a) One might start in such cases with the minimum form *atti* and go to the maximum form *khādati*[1] by adding one *a* in between. (b) One might also start, on the other hand, with the maximum form *khādati* and proceed to the minimum from *atti,* only to discover the absence of the infix *a* in between. The two statements according to the two approaches will be as follows :

(A) First, starting with the minimum *atti:*
atti = at + ti = ad + ti
khādati = khāda + ti
= *khād + a + ti*

(B) Second, starting with the maximum *khādati:*
khādati = khāda + ti = khād + a + ti
atti = at + ti = ad + ti = ad + o + ti.

It can be seen, therefore, that methodically the concept of 'zero' is as inevitable in the second kind of analysis as it is in the positional arithmetic. It is this second kind of analysis which has been in favour with modern structural linguistics. It is clear from Pāṇini's sūtra 2.4.72 (quoted above) that he too had favoured this approach in certain cases.

In all this process, a basic comparison between two forms is implied. In situations like the above, we have the following three alternatives open for description:—

1. to describe A with reference to or in comparison with B, or
2. to describe B with reference to or in comparison with A, or
3. to describe both independently of each other.

The third alternative obviously is just unscientific and entails *gaurava,* especially when the number of cases involved in description is very great. Out of the remaining alternatives, the second one is briefer and more scientific.

This brings into foreground one of the very important aspects of Pāṇinian technique of grammatical description. Pāṇini tries to go deep to a level at which the deep structures of all the entities he wants to describe can be fitted in a single formula which may serve as the starting point for the description of all—or, at least a large number of the cases. Of course, this is not the only method Pāṇini adopts invariably. He sometimes adopts a different technique as well. The type of the technique to be adopted is again based on statistical considerations. The technique which can cover the largest number of cases is taken to be the most preferable starting point from the point of view of brevity, comprehensiveness and convenience.

In the process of starting from the maximum common deep structure for the relevant forms and coming to the various types of minimum surface structures, Pāṇini has to omit the elements which were not available in the surface structures but which are assumed to be present in the deep structure. Thus, because the deep structure of *khādati* = *khād* + *a* + *ti* = $R + S_1 + S_2$, contains three morpheme-entities, the deep structure of *atti* also should contain three morpheme-entities, viz. *atti* = *ad* + *a* + *ti* = $R + S_1 + S_2$. But *a* is not available at the surface structure of *atti*; hence, *a* is zeroed; and we have, *atti* = *ad* + *a* + *ti* = $R + S_1 + S_2 = R + S_1 + S_2 - S_1$.

Such a method of addition and subtraction is many times used in mathematics in order to bring the given expression in line with the general formula of solution. Thus, to illustrate, take the mathematical expression $a^4 + 64b^4$ for factorisation. This expression does not fit in any general formula for factorisation; what is, therefore, done is to add a suitable term like $+ 16a^2 b^2$ and the expression takes the form $a^4 + 16a^2b^2 + 64b^4$.

But this expression is not identical with the original. The term $16a^2 b^2$ is, therefore, subtracted from it and the expression looks like,

$$a^4 + 16 a^2 b^2 + 64 b^4 - 16a^2b^2,$$

which now fits in the general formula of factorisation of difference of two squares; and the factors are —

$$(a^2 + 4ab + 8b^2) (a^2 - 4ab + 8b^2).$$

This exactly is what Pāṇini has done in the case of *ad* + *a* + *ti* – *a* in particular and in all other cases of linguistic zero in general. The form *atti* corresponds to the original mathematical expression, viz. $a^4 + 64b^4$.

In expanding *ad* + *ti* to *ad* + *śap* + *ti*, what Pāṇini seems to be aiming at is to bring the two-term expression *atti* to the level of a three-term expression *khādati*, because both the expressions belong to the category of the conjugated verbal forms. He could have divided all the *dhātus* into two separate categories, viz. (i) those with *vikaraṇas* and (ii) those without *vikaraṇas* and described the forms accordingly. But such a method would have assumed two categories of *dhātus* instead of one, which would have entailed a *gaurava*, as against *lāghava*, from technical point of view and it would have been perhaps rejected in preference to the former method. Assuming a single-structure formula of three terms like $R + S_1 + S_2$ for all the conjugated verbal forms obviously brings a natural symmetry in linguistic description, which is the most coveted principle in any scientific writing.

This method of description assures of greater brevity in the long run than the other one, viz., to describe A, the maximum, with a reference to or in comparison with B, the minimum. In this method of description, from minimum to maximum, there will always be only additions which are termed as '*āgamas*' in Pāṇini's grammar in every case and never any subtractions. The possibility of subtractions, that is zeros, is offered only in the description from maximum to minimum.

The technique of zero is thus basically and purely a technical device born out of necessity for brevity and symmetry of linguistic description and nothing else. As L. Bloomfield[2] put it, "The Hindus hit upon the apparently artificial but in practice eminently serviceable device of speaking of a zero element". To see in this technique, as S. Al George[3] does, an attempt on the part of Pāṇini to base his grammar on the principle of a sort of "ontological abstraction" and "conceiving the general, as having a richer comprehension than the subordinate individuals" is, from a purely technical point of view, to see too much in a grammatical work like Pāṇini's *Aṣṭādhyāyī*, whose aim is to describe a language and not to spell out any philosophy of language; the only technical principle that Pāṇini might have and has actually used is the principle of zero, perhaps borrowed from positional mathematics.

Examples from the Level of It-saṁjñā

One of the most widely used, and hence favourite, techniques of Pāṇini is the *It-Saṃjñā*, i.e. the technical term called '*it*'. The term is entirely devoid of any linguistic or social significance;[4] it conveys a purely technical meaning in Pāṇini's grammar. The term *'it'* is the name or label of the sound which Pāṇini defines as *'it'*. The conditions under which the sounds or even sound-groups are termed as *'it'* are well defined by Pāṇini himself in his grammar (cf. the sūtras 1.3.2-1.3.8 which are in all 7). The *it*-sounds are also called as *'anubandhas'* by later commentators and grammarians.

At the end of all these sūtras which lay down the conditions under which the different sounds or sound-graphs should be taken as *'it'*, there comes one of the most important sūtras in Pāṇini's grammar which states the purpose of laying down all these *'it*-sounds'. The sūtra is *tasya lopaḥ*, 1.3.9; it lays down that the sounds or sound-groups which are termed as *'it'* under conditions specified in the sūtras 1.3.2-1.3.8 disappear or amount to zero. Thus all the sounds or sound-groups which are termed as *'it'* in the preceding seven sūtras invariably amount to zero. Take, for example, the termination *ṣyañ* laid down by Pāṇini in the sūtra 4.1.78. As *ṣ* occupies the initial position of a termination, it will be termed as *'it'* according to the sūtra, *ṣaḥ pratyayasya*, 1.3.6 and will disappear according to *tasya lopaḥ*, 1.3.9; as the sound *'ñ'* a consonant, occupies the final position of a termination, it will be called as *'it'* according to sūtra, *halantyam*, 1.3.3 and is elided according to 1.3.9 quoted above. Now, the remaining sound-group viz. *-ya-* does not satisfy any condition of being *'it'* laid down in the sūtras 1.3.2.-1.3.8 and hence it does not amount to zero; it retains its form and existence.

The same is the case with the *it*-sounds in the *dhātus* or roots, *āgamas* or augments and *ādeśas* or replacing sounds or morphemes.

We find that in Pāṇini's grammar different terminations, *dhātus*, or *āgamas* or *adeśas* are accompanied by different *it*-sounds. Not only this but also the same termination or *dhātu* or *āgama* or *ādeśa* is found to have been accompanied by many *it-* sounds in different places and circumstances. Let us take the termination, say, *-ya* which has a wide distribution in the sphere of the *taddhita, kṛdanta*, verbal conjugation and transformation. This single termination is reconstructed by Pāṇini with different *it-* sounds in different places. The different reconstructions of this termination are: *ṣyaṅ*, *ṣyañ*, *yak, yat, yaṇ, yañ, kyac, kyaṅ* and *kyap*. It will be seen that the same termination *-ya-* is accompanied by as many as 8 different *it*-sounds. The same is the case with the *āgamas, ādeśas* and *dhātus*.

Just as we find the same termination or *dhātu* or *āgama* or *ādeśa* accompanied by different *it*-sounds, we also find the same *it*-sounds accompanying many terminations or *dhātus* or *āgamas* or *ādeśas*. Take, for example, the sound *c* which becomes *it* both in the initial as well as final position of a termination according to the *sūtras, cuṭū*, 1.3.7 and *hal antyam*, 1.3.3 respectively. The different terminations which have *c* as the *it*-sound are: *itac, ac, imanic, kānac, śānac, cānaś, khac, kyac* etc. So also with the *dhātus, gamḷ, śakḷ, vidḷ, mucḷ, sicḷ* etc. which have *ḷ* as their *it*-sound. So with the *āgamas* and *ādeśas*.

This is rather a strange technique — to assume something merely to negate it immediately. The *it*-sounds seem to be introduced with the aim to account chiefly for the difference in the formations. But what is the use of such differentiating agents which cease to function beyond differentiating from one another the terminations or *dhātus* or *āgamas* or *ādeśas* only? At this stage, such differentiating symbols require logically a positive, creative, operational capacity by which they will automatically operate and generate the formation. This would require special rules which would empower these differentiating symbols to operate, that is to say, to bring about the necessary phonological, accentual or any other type of modifications required to generate the desired formations. And this is exactly what Pāṇini has done. He has infused the *it*-sounds with differnt powers to bring about the required modifications in the grammatical process of the formation in question, laid down the rules and conditions, and specified the limit of the spheres in which they would exercise their powers.

The source of *it*-technique

As we have suggested earlier, each type of zero-technique used by Pāṇini in the three different spheres may have a different background, history and source. So is the case with the *it*-zero as well.

The *it*-sounds, as we have discussed earlier, are not linguistic units; they are extra-linguistic entities in so far as their purpose, function and nature are concerned. In nature, they are just means of the technique and their function is that of a sign, marker or symbol — whatever one may call them — which symbolises or stands for or represents some specific grammatical functions ascribed to them by Pāṇini in well-defined rules.

This technique of words or sounds symbolising, not a particular linguistic category but some process or entity, has its roots in the pre-Pāṇini Vedic literature. If we read the voluminous literature of the Vedas, the Brāhmaṇas, the Upaniṣads etc. we find they have profusely made use of the technique of symbolism of sounds to explain mythologies, parallelism, co-relations and/or correspondences of two entities on the same or on different planes.

We find the deities like *agni, indra* etc. symbolising some qualities or process or actions. The following are the references for the deities *agni* and *indra* :

For *agni: agniḥ prajanam* (symbolising function or action or process) G.B. 1.2.15; *agniḥ prajāpatiḥ* (symbolising a deity), KS, 22.7; 10; *agnir eva savitā* (deity), JU 4.12.1.1; *agnir eva brahma* (entity), G.B. 1.5.15., KS. 8.4; *agnir vā aśvamedhyasya yoniḥ*, TB. 3.9.21.2; *agnir vāva pavitram*, TB. 3.37.10; *agnir vāva yamaḥ;* TS 3.3.8.3; GB. 2.4.8; *agnir amṛtam* (symbolising something abstract), MSB. 10.2.6.17 etc.

For *indra : indro dyauḥ* (symbolising something abstract), *vy uta bhūmiḥ* (concrete entity), *indraḥ samudraḥ*, MS. 4.14.7; *indro mādhyandinaṃ savanam* (symbolising the process of pressing *Soma*); *indro yajamānaḥ* KS 7.10 etc.

The references of this type can be multiplied thousandfold. What these equations based on symbolism mean is that, apart from their physical correspondences, the symbols *agni, indra* etc. stand for the functions or qualities of the things they symbolise; or, vice versa, the functions which are symbolised by words like *indra, agni* etc. stand for the latter.

Not only this, but this technique of symbolism by words or phrases has been brought down to the level of sounds which are made to symbolise something, abstract or concrete. And this process began in the Vedic times itself. Examine, for example, the following single sounds functioning as symbols, used in the Vedic literature:

a iti brahma, A.Ā, *2.3.8; akāro vai sarvā vāk* AĀ *2.3.6; ā prati ca vā sarvam,* JB i.180.
om iti vai sāma, om iti manaḥ, om iti indraḥ, JU. 1.2.2.2;
om iti vai svargo lokaḥ. AB. 5.32; *rasa oṃkāraḥ,* JB. 2.78. etc.
kam vai prajāpatiḥ, Kāṭhakasaṃkalana, 48.1; 138.5; *prajāpatir vai kaḥ,* TS 1.6.8.5. etc.

It is on the basis of this symbolism of *ka* = Prajāpati, that the famous Ṛgvedic passage, *kasmai devāya haviṣā vidhema*, ṚV. 10.121.1-9 is explained as *prajāpataya devāya haviṣā vidhema*, by substituting the word *prajāpatī* for *ka* in the stanza.

chidraṃ kham ity utkam, GB. 2.2.5
ṇakāro balaṃ ṣakārah prāṇa ātmā, AĀ. 3.2.6;
om iti satyam, na ity anṛtam, AĀ. 2.3.5.

Innumerable instances of sounds standing as symbols for some god or thing or entity (abstract or concrete), function, action or process, qualities etc. or for anything under the sun for that matter, can be quoted from pre-Pāṇinian literature. The *ekākṣarakośa* (= dictionary of single sounds) in Sanskrit is an example of this kind of single sounds having meaning. Even BD quotes two single-sound *nipātas* viz. *i* and *u* having a meaning; cf. BD on the Pāṇini, 1.1.14 (*nipāta ekājanāṅ*): *i vismaye, i indraḥ, u vatarke, u umeśaḥ*; these symbols vary from single sounds to a group. The Tantra literature in India and Tibet is a good example in this respect which uses symbols like *am, kam, jam, vam, lam* etc. as also mystic utterances like *hrām, hrīm, hrūm, hrom* etc. which are apparently meaningless but which, according to the tradition of the tantras, have a great significance and convey a definite meaning. In all the above examples of sound-symbols, the symbols are used as substitutes for the names or qualities or functions of the deity concerned. As soon as the deity appears or manifests or functions, the symbols are discarded. They are used as symbolic abbreviations for the names of the deity.

With this background, it will be easy to see that what Pāṇini has done in employing the sounds as symbols for some grammatical function is exactly the same as that of the tāntrikas or Vedic seers above. He also takes the single sounds as symbols, terms them as *it*, invests them with definite powers in pre-defined spheres and lets them loose to work; and they work accordingly. But he goes a step further than the Vedic seers; he also invests the sounds with powers of symbolising grammatical functions and processes; and not merely of differentiation.

The *it*-symbols can also be compared to what is called an *upalakṣaṇa* (lit. subordinate or secondary adjective) in Nyāya. It is contrasted with the *mukhya lakṣaṇa* called *viśeṣaṇa*. The *viśeṣaṇa*, which is a permanent mark of something, is always associated with the thing, as in the example, *daṇḍī saṃnyāsī*; the *daṇḍa* is a permanent marker of *saṃnyāsa*; whoever is a *saṃnyāsī* must keep with himself the *daṇḍa*. The *upalakṣaṇa* is a temporary, not permanent, mark of a thing, as in the example, *kākavad devadattasya gṛham*, 'The house on which the crow is presently sitting belongs to Devadatta'. Here, the crow may fly away; yet the house remains as and where it is and still belongs to Devadatta.

In the same way, the *it*-sounds are the *upalakṣaṇa* of the grammatical modifications; the grammatical modifications brought about by the *it*-sounds stay on, though the *it*-sounds themselves are made to 'fly away' like the crow.

Examples from the Sphere of Anuvṛtti

1. There is yet another sphere in the grammar in which Pāṇini has very skilfully employed the technique of zero which has formally a position. The peculiarity of the technique of zero in this sphere is that it is nowhere stated explicitly in clear terms nor is it implicitly suggested nor again is it stated to be amounting to zero indirectly. Yet, one so easily and automatically takes the zero of the categories concerned for granted that the zero employed in this technique goes totally unnoticed by the reader of the sūtras. The zero in this sphere relates solely to the field of the technique or principle of sūtra-arrangement.

As is well-known, one of the principles of the arrangement and interpretation of the sūtras in the *Aṣṭādhyāyī* is what traditionally goes by the name *anuvṛtti*,[5] lit. following, pursuing etc. (from *anu* + *vṛt*), that is, borrowing, carrying over etc. In simple language, if two sūtras have certain words/conditions/categories common between them, the succeeding sūtra drops out or does not repeat the common words of the preceding sūtra and yet the common words are to be automatically taken to be present in the succeeding sūtra. To illustrate, let us take the two sūtras, *upadeśe aj aunāsika it*, 1.3.2 and *halantyam*, 1.3.3. While interpreting the succeeding sūtra *halantyam*, 1.3.3, the words *upadeśe* and *it* from 1.3.2 are to be taken as being present in it and the sūtra 1.3.3, when expanded by reading the two words will read — *upadeśe halantyam it*. While writing the grammar Pāṇini includes the common words *upadeśe* and *it* in the preceding 1.3.2 but drops them out in the succeeding 1.3.3. This device is employed sheerly out of the consideration for axiomatic brevity in the sūtra arrangement; the brevity thus achieved by this technique is *alpākṣaratva*,[6] 'having less number of letters or sounds'.

Viewed from the point of view of a comparison of wording of the two sūtras, this technique of dropping out the common words from the succeeding sūtras, yet at the same time, making readers believe their presence in the same, amounts indirectly to the zeroing of the words, or in other words, employing indirectly the technique of zero of common elements. This device is similar to the device of taking out common factors in the factorisation process in mathematics.

The technique of *anuvṛtti* in greater detail

Let us take, as an example, the sūtras 1.3.2 – 1.3.8, in all seven, laying down the *it*-sounds and the conditions for the sounds to be *it*. The following is the Pāṇinian statement and order of the sūtras which are quoted here for easier and clear understanding :—

(a) Pāṇinian statement :

1. *upadeśe ac anunāsika it,* 1.3.2
2. *halantyam*, 1.3.3
3. *na vibhaktau tusmāḥ*, 1.3.4
4. *ādir ñiṭuḍvaḥ*, 1.3.5
5. *ṣaḥ pratyayasya*, 1.3.6
6. *cuṭū*, 1.3.7
7. *lasakavataddhite*, 1.3.8.

All of these lay down the different conditions for the various sounds to be *it*; sūtra No. 3, viz. 1.3.4, however, prohibits certain sounds to be *it* under certain conditions.

If we read the sūtras, we find that except the one, single sūtra, viz. 1.3.2, which contains the words *upadeśe* and *it*, all the rest do not have any such words; yet, we have to borrow these words in consistently interpreting the rest of the sūtras. While sūtras Nos. 2, 3 and 4 viz. 1.3.3, 1.3.4 and 1.3.5 require the *anuvṛtti* of two terms viz. *upadeśe it*, sūtra No. 5, viz. 1.3.6, *ṣaḥ pratyasyasya*, requires the *anuvṛtti* of *upadeśe it* (from the first sūtra, i.e. 1.3.2) and *ādiḥ* (from the fourth sūtra, i.e., 1.3.5); sūtras Nos. 6 and 7 viz. 1.3.7 and 1.3.8 on the other hand are meaningless without the *anuvṛtti* of four terms viz. *upadeśe, it, ādiḥ* and *pratyayasya* (the last from the fifth *sūtra* i.e. 1.3.6). Not only this, but also that the absence of *anuvṛtti* of these terms in places where they are required will create chaos and upset the whole Pāṇinian sūtra-system; its absence will hinder any intelligent interpretation of the sūtras, will create technical and grammatical difficulties and further lead to interpretational deadlocks, affecting the grammar of the language. The importance of *anuvṛtti* in the Pāṇinian principles of axiomatics can, therefore, hardly be exaggerated.

We know, therefore, that many times and in many places in Pāṇini's sūtra-system, we have to read many sūtras by assuming the existence of certain words even if they are not there. This assumption of their existence there is based upon their actual inclusion in some preceding sūtra or sūtras. If, therefore, in the above example, we put the borrowed words in the places in which they are required but are not there, we may perhaps get a clue as to the underlying principle of *anuvṛtti*. The Pāṇinian statements in their solved form will then be :—

(b) Solved Pāṇinian statements :

1. *upadesé ac anunāsika it*, 1.3.2
2. *upadeśe halantyam it*, 1.3.3
3. *upadeśe na vibhaktau tusmā antyam it*, 1.3.4
4. *upadeśe ādir ñiṭuḍavaḥ it*, 1.3.5
5. *upadeśe ādiḥ ṣaḥ pratyayasya it*, 1.3.6
6. *upadeśe ādiḥ cuṭū pratyayasya it*, 1.3.7
7. *upadeśe ādiḥ laśakvataddhite pratyayasya it*, 1.3.8

The dissolved form of the Pāṇinian statements at once now helps us to decipher the principle on which the technique of *anuvṛtti* seems to be employed. We can easily see that the words, which are required to be borrowed are common words in all or some of the Pāṇinian statements. In other words, what Pāṇini did is that he took out all the common words occurring in all the sūtras and inserted them in one single sūtra which would stand at the head of all its dependent *sūtras*. The concept underlying the principle of *anuvṛtti* is, therefore, purely mathematical, viz. that of taking out of the brackets the highest common factor from amongst the given, numerous algebraic expressions. The Pāṇinian statement of rules, therefore, looks like bracketed algebraic or mathematical expressions with the highest common factor or factors out of the brackets. The point will be clear if we put symbols for the words and rewrite the sūtras symbolically.

Put *a* for the word *upadeśe*
b for the word *ac*
c for the word *anunāsika*
d for the word *it*
x for the word *hal*
y for the word *antyam*
p for the word *na vibhaktau tusmāḥ*
z for the word *ādiḥ*

m for the word *ñiṭūḍayaḥ*
n for the word *ṣaḥ*
q for the word *pratyayasya*
k for the word *cuṭū*
w for the word *lasakvataddhite*;

and the sign + (plus) for showing the Pāṇinian order or sūtras. We now represent the solved Pāṇinian statements symbolically as : $(a.b.c.d) + (a.x.y.d) + (a.p.y.d) + (a.z.m.d) + (a.z.n.q.d) + (a.z.k.q.d) + (a.z.w.q.d)$.

Taking out the common factors out of the brackets, we get, $ad \times b.c. + y\ (x + p) + z\ \{m + q\ (n + k + w)\}$ which is the perfect algebraic representation of the Pāṇinian statement of the seven sūtras. While the factors *xy* are common to all of the sūtras (and hence they are taken out of the biggest brackets), the factors *y*, *q* and *z* are only partially common to some of the sūtras. The principle followed by Pāṇini in employing the technique of *anuvṛtti* in his sūtra-system, therefore, seems to be similar to the simple principle of factorisation or finding out the common factors from amongst the given expressions in algebra. The common factors taken out may be a single word or group of many words. And to put in another way, the mathematical technique of finding common factors seems to be based on the principle of *anuvṛtti*.

Another point of view

The method of factorisation shown above being adopted by Pāṇini as the basis of the *anuvṛtti*-technique is one way of looking at it. There is yet another way in which we can arrange the sūtras in terms of categories and not words which brings out clearly the use of indirect zero in the technique. It is as in the following. In all there are seven categories in the *anuvṛtti*-technique adopted in the case of the *it-saṁjñā* sūtras (see the table on the next page).

The categories or elements which are to be carried over in the succeeding sūtras, are dropped out by Pāṇini; in other words, they are zeroed; hence they are represented by zero in the above table. The entities which are not to be carried over in the following sūtras are represented simply by a horizontal dash.

It will be clear from the above way of writing the sūtras that what Pāṇini does in adopting the technique of *anuvṛtti* is that the first statement (in this case the sūtra, 1.3.2) from which he starts employing the technique

Technical condition	Sounds which are it	Phonological condition	Morphological condition	Positional condition	Negation	Result	Sūtra No.
upadeśo	*ac*	*anunāsaki*	—	—	—	it	1.3.2
0	*hal*	—	—	*antyam*	—	0	1.3.3
0	*tusmāḥ*	—	*vibhaktau*	0	*na*	0	1.3.4
0	*ñiṭuḍ-avaḥ*	—	—	*ādiḥ*	—	0	1.3.5
0	*ṣaḥ*	—	*pratyayasya*	0	—	0	1.3.6
0	*cuṭo*	—	0	0	—	0	1.3.7
0	*lasaku*	—	0 *and ataddhite*	0	—	0	1.3.8

is extended to or contains maximum number of categories or terms and then by eliminating or eliding the common terms which need not be repeated he frames the wording of the succeeding sūtras. In this case also, as in the case of the linguistic zero discussed before, he starts with maximum expanded form of the sūtras and then goes to the minimum statements. It will also be clear that out of the 49 cells (7 columns and 7 rows), Pāṇini has used the zero in as many as 18 cells of categories, 16 cells are completely vacant and only 15 cells are filled up with positive categories. The zero-cells thus form a majority number in this case.

There are many examples in Pāṇini's grammar in which the technique of *anuvṛtti* is employed. The scope of *anuvṛtti* varies from two to a number of sūtras: cf. the sūtras 3.2.95 and 3.2.96 where the *anuvṛtti* of the word *yudhikṛñaḥ* is limited only to the two sūtras, whereas the *anuvṛtti* of the word *ṭac* extends up to 22 sūtras from 5.4.91-5.4.112; while the *anuvṛtti* of the word *dhātoḥ* in the sūtra, 3.1.91 covers almost one, entire *adhyāya* (viz. the 3rd) of four *pādas* containing in all 541 sūtras, the *anuvṛtti* of the word *prātipadikāt* in the sūtra 4.1.1 embraces in all 1,190 sūtras from two complete *adhyāyas* (viz. 4th and 5th) of four *pādas* each.

The *Anuvṛtti*-zero—Its Source

When we come to search for the source of the third type of zero-technique used by Pāṇini in the sūtra-arrangement in order to achieve brevity, we have to search for a different field. The data, of course, are the same, viz. the pre-Pāṇinian Vedic literature, but with a difference.

As we know, the Vedas are analyzed into their different words, morphemes and suffixes by the *padapāṭha*, obviously with a view that there should not be any ambiguity as regards the Vedic meaning based on different analyses by different people. The *padapāṭha* of the ṚV is done by Śākalya traditionally. The *Padapāṭha* is, therefore, rightly called as 'a commentary'[7] on the Vedas, perhaps the first commentary attempting the explanation of the Vedic verses. While writing the *Padapāṭha*, the *padapāṭhakāras* confronted a peculiar situation in which repetitions of whole verses, verse-feet and even of words were involved.[8] And the problem before them was : whether all these repetitions of the verses, partially or fully, were to be subjected to re-writing or re-reciting? That would have been a gigantically onerous job. They, therefore, ingeniously devised a technique to avoid and drop the repetitions of the former passages in successive passages. This technique is known as the technique of *galita* or *galanta* (lit. dropped down) and the dropped-out passages were named as *galitas*. The term *galita/galanta* is quite well-known in the circles of Veda-reciters. It is defined as: *padapāṭhe kadācit pūrvaprāptānāṃ padānām punaḥprāptyavasare pāṭhapradarśanaparityāgaḥ/ sa parityāgaḥ samayaśabdena vyavahṛyate/ parityaktāni padāny api tena śabdenocyante/*[9] What this means in short is that the repeated *padas* are dropped out and this dropping out is called by the name *samaya*.[10] The rules of accepting this or that as *galita* vary from Veda to Veda.[11] But that all the Vedas adopted this or that system of *galita* in the *Padapāṭha* is an irrefutable fact. The words which were thus dropped were not repeated again by the reciters of the Vedic *Padapāṭha* as well as of the Vedic *Saṃhitā* and were marked in the old manuscripts by the zero sign or by the word marked *galita* itself.[12] Thus, ṚV. 2.1.16 is wholly repeated at 2.2.13; the latter *ṛc* is, therefore, the *galita* and is represented by zeros or the word *galita*; cf. also the *ṛc* 2.11.21 repeated at 2.15.10, 2.16.9, 2.17.9, 2.18.9, 2.19.9 and 2.20.9 at which places it is marked by zero.

That the *galitas* were dropped out from the succeeding *Padapāṭha* does not mean that they were not part of the *Saṃhitā*-text at that place. They enjoyed in the *Saṃhitā* the same status as others which were not *galitas* and were recited by some in their proper places in which they originally occurred.

Now, given this fact and phenomenon in pre-Pāṇinian times regarding the technique of dropping out from the succeeding part the identical or repeated portion as a means obviously of brevity and saving time, we immediately come to know the origin and source of the Pāṇinian technique of dropping out the identical words from the successive sūtras. Pāṇini, as a skilled technician, takes the hint and employs the technique in the arrangement of the sūtra. And we have this wonderful technique which we call by the name *anuvṛtti*.

As a general rule, the condition for dropping the repeated words in the *Padapāṭha* is that the number of words repeated must always be more than three or be a whole *pāda* or verse.[13] But Pāṇini narrows down this condition to a single word or even letter. The obvious reason is that Pāṇini's grammar being purely formal is not at all concerned with non-formal category of meaning. For Pāṇini, a simple, single sound is fully on par with a full big sentence for purposes of elision on the formal plane. The *padapāṭhakāra*, however, is concerned with right divisions and elision of *pādas* which convey meaning.

COMPARISION OF PAṆINIAN ZERO WITH THAT IN OTHER SCIENCES

We have seen before the working of the Pāṇinian technique of *lopa* or zero on both the technical (i.e. *it-saṁjñā* and *anuvṛtti*) as well as on non-technical levels (i.e. linguistic/grammatical description). We have also tried to trace a possible origin of the technique in pre-Pāṇinian Sanskrit literature of the Vedas, the Brāḥmaṇas and the Upaniṣads.

But this is only a partial view of the technique. In developed civilizations, as knowledge expands and advances in all fields, there is always the inevitable stage at which the phenomenon of give-and-take, of borrowing-and-lending of methods, techniques and results of one science by the other occurs. Sanskrit civilization is also no exception to this. The Pāṇinian technique of zero cannot be — and, as we have seen, is not — a sudden offshoot of its times. Just as it has evolved out of the pre-Pāṇinian methods and techniques of scientific investigation, it is also likely that it might have been borrowed from other sciences or that it might have been adopted by other sciences. To ascertain whether the technique is borrowed from or is lent to other sciences, it is necessary to make a search for any parallels of this technique in other sciences. If we look for possible parallels of the technique in other sciences, we find that the sciences which present or employ a technique parallel to the Pāṇinian zero are three. They are: the Buddhistic philosophy of *Śūnyavāda*; the Nyāya-Vaiśeṣika concept of *abhāva* and the technique of zero in mathematics. It will, therefore, be very interesting to compare and contrast all the aspects of zero from all these three sciences.

Buddhistic Concept of Śūnyavāda

It was first expounded by Nāgasena in his conversation with King Milinda (who is generally identified with the Greek King Minander of Ghazni and Kabul). Put in nutshell, Nāgasena's argument is as follows: If we see a thing, we perceive after analysis that the thing is made up of various parts. None of the parts can be fixed as 'the thing'. If, therefore, we separate the parts one by one, what we get ultimately is that the thing vanishes and there is a total vacuum. The same is the case with the universe. If we take away all the elements which go to make up the universe and try to see its root-cause, what we find in ultimate analysis is that there is no root-cause at all at the basis of the universe;[14] there is total vacuum.

Empirically speaking, there is no regular procedure laid down by Nāgasena in arriving at this type of Śūnya as the basis of the universe. The whole argument seems to be based on simple 'negation' of all the parts which the thing is made up of. In the absence of any logical and scientific procedure, it is difficult to digest the *Śūnyavāda*. Secondly, the Buddhistic *śūnya* is not a technique but a philosophical theory. As such, it cannot be compared with the Pāṇinian zero, which is a technique.

Naiyāyika's Abhāva

The next science which discusses the zero is the Nyāya Vaiśeṣika. The zero in Nyāya is termed as *abhāva*, which literally means 'absence of *bhāva*, i.e. existence (*a-bhāva*)'. Out of the many definitions of *abhāva* which the Nyāya-system proposes, the most agreeable are only two. They are:—

1. *bhāvabhinnatvaṃ abhāvattvam*, which means 'the quality of being different (i.e. opposite) from existence is *abhāva*; and
2. *pratiyogijñānādhīnajñānavattvaṃ abhāvatvam*, which means 'the counterpositive is cognised in the cognition of the absence and. . . the cognition of the absence is dependent upon a cognition of the counter correlate.'[15]

Out of the two, the second definition is more acceptable. In simple words, what the definition means is: $+ x$ is the entity, which is the opposite (*pratiyogin*) of $-x$. The knowledge of $-x$ depends upon the knowledge of $+ x$. Thus, the entity $-x$ indicates the *abhāvatva* of $+ x$; or vice versa, $+ x$ indicates the *abhāvatva* of $-x$.

This is perfectly all right. But this simply means 'the absence or negation' of a thing; that is, it only gives the knowledge of 'the negative or minus' side; it does not necessarily mean 'zero' since it only leads to comprehend the idea of 'opposite'. It cannot be a theory of zero since it does not lay down an actual positive zero; it is also not a technique. Matilal rightly calls it as 'a doctrine of negation' and not a theory or technique of 'zero'. It is thus a simple negation and not a positive concept.

Abhāva is reckoned for the first time as an independent category by the Vaiśeṣikas after Praśastapāda. Then we find it in Udayana's *Kiraṇāvali*. Śivāditya's *Saptapadārthī* includes it as the seventh *padārtha*.

Mathematical Zero

When we come to mathematics, we face two types of zero: (i) as a number as 0, 1, 2, 3 . . . etc. and (ii) as the symbol for a vacant variable, i.e. place. As a number in the series 0, 1, 2, 3 . . . it is "an immediate predecessor" of the positive number 1; as a value of a place-variable, it indicates an empty space in numbers like 10, 20, 30 etc. The zero as a number is a mathematical concept and is totally different from the symbol '0' which is used to indicate an empty, unoccupied place in numbers like 10, 20, 30 etc. The former zero, viz. the number can be obtained by the mathematical operation of subtraction of equal entities, viz. 1-1 or in general x-x. The latter zero, viz. the symbol of a vacant place cannot be comprehended without the place-notation.

Out of the two types of zero, the zero as a number, preceding the positive integer 1, can be roughly compared with the *śūnya* of the Buddhistic philosopher Nāgasena. Just as Nāgasena arrives at zero by negating or subtracting all the parts of a thing, the mathematical zero can also be arrived at by subtracting a quantity from itself. To illustrate by the example of a 'chariot' taken by Nāgasena,

Chariot = the seat + the canopy + the yoke + the wheels etc. Putting symbols like x, y etc. for the different parts, we have the equation —

C(hariot) = $x + y + z + a + b + c$ . . . etc.

If now we subtract one by one all the parts viz. x, y etc. we get,

C = ($x + y + z + a + b + c$ . . . etc.) – ($x + y + z + a + b$ etc.) = 0.

The same procedure is visible in getting the number zero in mathematics. To illustrate, the number 0 = 1–1, or generally, x–x.

The number zero in mathematics can thus be easily compared with the *śūnya* of the Buddhistic philosophy of *śūnyavāda*, since the procedures to arrive at *śūnya*/zero are the same in both the sciences. The mathematical zero as a marker for a vacant place in numbers like 10, 20, 30 etc. however, cannot be compared with the Buddhistic *śūnya*, since the Buddhist *śūnyavāda* has no procedure similar to the one in mathematics in this case.

Let us now examine the Pāṇinian technique of zero used in linguistic description of the verbal forms *khādati* and *atti*. *Khādati* = *khad* + *a* + *ti*, i.e. $R + S_1 + S_2$, and therefore, *atti* = $R + S_1 + S_2$ = *ad* + *a* + *ti* = *ad* + *o* + *ti* (vide the procedure given before).

The main principle underlying such an analysis is: since *khādati* as a verbal form is composed of three elements viz. $R + S_1 + S_2$, *atti* as a verbal form must also be represented as composed of three elements viz. $R + S_1 + S_2$; but since the second element s_1 is not available in *atti*, it is zeroed.

Let us now examine the underlying principle in mathematics in representing 10 as a two-digit number with the symbol of zero.[17] We have a word like *pañca-daśa* (symbolized as 15) and a word like *daśa* (symbolized as 10).[18]

Now comparing the two number-words, we find:

1. The word *pañca-daśa* is linguistically made up of two linguistic morphemes viz. *pañca* and *daśa*.
2. The word *daśa* exhibits only one morpheme and has no other morpheme.
3. The word *daśa* is common between the two words; therefore,
4. the two being number words can be easily compared and set in only one kind of symbolic representation.

Let us now put numerical symbols for the two words *pañca* and *daśa* as 5 and 1 respectively; and we have the symbolic representation[17] as:

pañca-daśa = 51 = 15; and
-daśa = 1 = 01 = 10

This procedure is exactly similar to the one followed by Pāṇini. The main principle seems to be to invent a general formula to fit in it all types of forms of the same kind for the sake of symmetry of description. Pāṇini on his part assumes three-morpheme structure for the verbal forms as basic and describes other structures of less morphemes with a reference to it. Mathematics on its part describes all numerical structures with a reference to an assumed structure which contains higher number of places; *pañca-daśa* has two words and hence two places; *daśa* has one word and one place; the other vacant place in *daśa* is filled up by the symbol zero. At this stage, therefore, we find that the techniques of description adopted by mathematics and Pāṇini are basically the same.

The underlying principles of the two different types of zero in mathematics, therefore, seem to go back to two different sciences. The principle behind getting the zero as a number by the procedure of subtraction of equal entities, viz. $x-x$ is similar to the procedure adopted by the Buddhistic *śūnyavāda*, if the mathematical procedure of subtraction conjectured and presented here is what was intended by Nāgasena when he negated all the elements which go to make up the universe. The principle behind getting the zero as a symbol for vacant place-value as in 10, 20 etc. is similar to the one adopted by Pāṇini who assumes structures with more places/morphemes as the basis.

We do not know who borrowed from whom.

RESUMÉ

1. We are now in a position to take a panoramic view of all the zeros used by Pāṇini in his grammar for the purposes of linguistic description of the Sanskrit language.

There are, as we have seen before, three types of circumstances in which the technique of zero can be said to have been employed by Pāṇini. The first type relates to purely linguistic description and the zero in this sphere can be rightly termed as 'linguistic zero' in the sense that it is employed for the sake of symmetry in description. This urge for symmetry presupposes a certain principle like structural approach underlying it. This type of zero can be suitably compared with the zero in mathematics, which is used to indicate a vacant place, as in 125 and 105 in which the zero in the latter indicates the place of decimal and can compare with the decimal place-number 2 in the former. Though numerically zero is not equal to or identical with the number 2, from the point of view of its capacity as an indicator of the place, it is fully on par with the latter. In the same way, though phonetically the *śap* = 0, it is on par with non-zero counterpart so far as its place-value is concerned.

Unfortunately, we do not get any pre-Pāṇinian work, either on mathematics or any other subject which uses this type of positional analysis and to which the Pāṇinian technique can be traced back. We have therefore, to remain satisfied with only its comparison with the post-Pāṇinian mathematical technique based on position, in which the number-word *daśa* is written as a two-digit symbol with zero, as 10.

2. When we come to examine the technique of *it-saṃjñā*, in which the zero of the extra-linguistic *it*-sounds is employed by Pāṇini for fruitful purposes, we enter into a purely technical field, dealing with the technique of description, not only linguistic but of any kind. The technique of *it-saṃjñā* or some such similar technique can be employed in any field by an intelligent scientist. The principle involved in this technique is of symbolism. The presence and absence of the *it*-sounds symbolise different grammatical functions in Pāṇini's grammar.

As we have said before, the technique of symbolising something by a briefer symbol is not unknown to the pre-Pāṇinian literature of the Vedas, Brāhmaṇas and Upaniṣads. This literature, therefore, seems to be source for the Pāṇinian technique of *it-saṃjña*.

We can analyse the technique of symbolism into four stages:

2.1. The pre-Pāṇinian stage of the Veda, Brāhmaṇa and Upaniṣads. In this stage, as seen earlier, the different deities, sounds, mantras stand for either some action (as in *agniḥ prajananam*) or concept (*om iti brahma*).

2.2 The Māheśvara–sūtras come second in line in pre-Pāṇinian times to utilise the technique of symbolising something with any — preferably shorter — phrase. The author of these sūtras employed the symbolism of the sound to indicate the final limit of the *pratyāhāra* and consequently to frame a *pratyāhāra* from the given sounds. This technique seems to have been borrowed by later mathematics in which the range or extent of a straight line is indicated by the initial and final points. The function of this technique is, however, limited only to this extent and the *it*-sounds of the Śiva-sūtras are not infused with any other power.

2.3 When we come to the Pāṇinian technique of *it-saṃjñā*, which is the third stage in the development of the technique of symbolism, we find that apart from taking hints from the Vedic literature and Śiva-sūtras, Pāṇini extended it to a still more complicated status of empowering the *it*-sounds with functions of different and varied grammatical operations of *vṛddhi*, *guṇa*, their absence, accent etc. And, as we have examined it, he has succeeded in it to an infinite extent, so much so that not only that the technique of *it-saṃjñā* does not look alien to the grammar, but that it creates an impression that it is the most natural to it and that it is the only technique by which the grammatical phenomena can be explained.

The third feature of the Pāṇinian technique is that it restricts its symbolism only to the formal plane of grammatical operations.

2.4. When we come to Kātyāyana we find that his technique of *it*-sounds not only transcends the limits of formality but it works exclusively on semantic plane like synonyms, meaning general (*sāmānya*) and particular (*viśeṣa*) etc. It thus presents a complete contrast to the Pāṇinian technique. One must, of course, admit that the technique of symbolism by sounds can be easily extended to the semantic plane and that Kātyāyana, borrowing perhaps from pre-Pāṇinian, Vedic and Pāṇinian texts, has certainly displayed a further stage of development.

This thus seems to be the journey of the technique of symbolism in general and sound-symbolism in particular — from concrete to abstract, from formal to non-formal, from simple to complex. The Sanskrit grammarians in particular, however, seem to have utilized the technique of the *it*-symbols and their consequent zero very widely, intelligently and successfully.

3. The third type of zero, viz. *anuvṛtti*-zero, seems to have been borrowed by Pāṇini from the *padapāṭhakāras* who use the technique of *galitas* or *galantas* in cases where the pādas, verses, distichs of the Vedic mantras are repeated in toto and the repeated part/s are indicated by zeros in the *Padapāṭha*. This same principle of dropping the common elements from the prior portion in the next, consequent portion is used, as we have shown in the mathematical procedure of factorization.

4. Coming to the points of comparison between Pāṇini and other three sciences of Buddhistic *śūnyavāda*, Naiyayika's *abhāva* and mathematical zero, one can only say that Pāṇini's zero can only be compared with the mathematical zero. The Buddhistic *śūnyavāda* is only a philosophical theory; the method by which the *śūnya* was discovered by Nāgasena is neither empirical nor very clear and does not stand the rigorous test of scientific investigation. The *abhāva* of the Nyāya-Vaiśeṣika system is only a concept; it is neither the philosophical theory nor a technique. It, therefore, cannot be compared with the Pāṇinian zero. The mathematical zero, however, presents a very good parallel to the Pāṇinian zero.

The points of similarity between the two techniques of zero employed by the two systems, viz. of Pāṇini and mathematics are as follows:

1. In both of them, the concept of zero attains the status of a technical device to build up the respective systems.
2. In both of them, zero represents the zeroed entity as well as its place.
3. Both presuppose positional analysis and representation.
4. In both of them, the zero of an entity is available as the counterpart of its non-zero entity in some other structures. This implies that the basis of employing zero as a technical device is comparison of two entities — one with zero and the other with non-zero elements.
5. In both of them, the concept of zero is not merely intellectual but can be comprehended, arrived at and represented by a definite empirical, formal process laid down in those sciences.
6. In both of them, the structure with non-zero elements contains greater number of positive elements than the one with zero elements.
7. In both of them, the zero requires to be inserted at the deep-structure level.

REFERENCES

1. The statement should be understood only with reference to the present two cases. There are still larger forms like *corayati* (10th conj.) or *jigamiṣati* (des. of *gam* 'to go') than the present *khādati*.
2. Cf. L. Bloomfield, *Language*, 1961, p. 132.
3. Cf. S. Al George, 'The Semiosis of Zero according to Pāṇini', *East and West*, 17, Nos. 1-2, Rome, 1967, pp. 116f., 119.
4. The term *it* is derived from *i̥*, 'to go', with the Pāṇinian zero-termination *kvip*; thus it would mean 'that which goes away'; cf. *Cāndravṛtti*, a commentary on the *Chādra Sūtra* 1.1.5: *eti gacchati iti it/ iṇaḥ kvip, tuk, etīti it/ gacchati, na tiṣṭhati lupyata ity arthaḥ.* Jainendra and Śākaṭāyana explain it as *aprayogī it* 'that which is not used in the language'. According to some scholars, '*it*' is '*iti*' itself with the final *i* of the latter lost; cf. G.B. Palsule, *The Sanskrit Dhātupāṭhas — A Critical Study*, University of Poona, 1961, p. 62. On the contrary, is it not more natural to say that *iti* is loc, sing. of *it*

with an accent on the final *i* as *ití* (cf. Pāṇ. sūtra, 6.1.168) and with an accent-shift as *íti*, it would be turned into an *avyaya*? Examples of regular declined words turning into an *avyaya* with an accent-shift are available in Vedas; cf. *divā*, mstr. sing. (ṚV 1.161.14 etc.) turning into an *avyaya* as *dívā* (ṚV 1.24.10, 12).

5. For *anuvṛtti*, cf. S.D. Joshī and Saroja Bhate, *The Fundamental of Anuvṛtti*, CASS, University of Poona, Pune 1984. The book is nothing but a collection of all possible situations in the *Aṣṭādhyāyī* in which the principle of *anuvṛtti/ maṇḍūkapluti* can be said to be present traditionally; it, however, does not spell out any 'fundamental' of *anuvṛtti*-technique, as its title claims.
6. Cf. the definition of a *sūtra* as : *alpākṣaraṃ asaṃdigdhaṃ sāravad viśvatomukham astobham anavadyam ca sūtraṃ sūtravido viduḥ.*
7. Cf. *Samayānukramaṇī*, Appendix IV, p. cix in *Ṛgvedānukramaṇī*, ed. C.K. Raja, Madras, 1932 : *saptamy anukramaṇikā śākalyam anudhāvati/ vyākhyātaivaṃ hi teneyam saṃhitā padadarśanāt / /*
8. For Vedic repetitions, cf. M. Bloomfield, *Ṛgveda Repetitions*, HOS. 20, 1916; also *Vedic Concordance*, HOS 10; 1906; reprinted in 1961.
9. Cf. *Ṛgvedaprātiśākhya*, ed. Pashupati Nath Sastri, Calcutta, 1927, p. 143.
10. Cf. Lakṣmīdhara Sūri, *Galitapradīpa*, ed. by Shri Shrikṛṣṇadeva, *Sarasvatī Suṣamā*, Saṃvat 2016, verse 2, p. 2: *alakṣyatvād anūhyatvād adṛṣṭānuvidhānataḥ/ apy eṣām prātiśākhyādyair ataḥ samayanāmatā//* He defines *galita* as: *brahmayajñapārāyaṇādiṣu prāthamyena proccyamānasya ṛcātmakavedasya dvitīyaprakṛtibhūte padapāṭhe prāyaśaḥ punaruktapadasamūhānāṃ lopo bhavati/ tasyaivātra galitasaṃjñā / /*
11. Cf. *Galitapradīpa, op. cit. prastāvanā, p. kha : paraṃ ceyaṃ galitapadavyavasthā sarveṣu vedeṣu bhinnā.*
12. For a more detailed study of *galitas*, cf. K.P. Jog, 'On the galitas in the Padapāṭha as a means of Ṛgvedic Interpretation', IA, Vol. II, Nos. 1-4, 1969, pp. 56-82; cf. also K.V. Abhyankar, *Vedapāṭhacarcā* with *upalekhāsūtram*, BORI, 1974.
13. Cf. *Galitapradīpa, op. cit.* p. kha, *sarvatra padatrayā nyūnam, galitam nāsti; punaruktaḥ sampūrṇo mantro galito bhavati.*
14. For detailed discussion, cf. M.D. Pandit, 'Zero in Panini' pp. 136-45.
15. Cf. B.K. Matilal, *The Navya-Nyāya Doctrine of Negation*, Harvard University Press, 1968. The word *pratiyogin* in the context of *abhāva*, according to Matilal means '*abhāva-pratiyogin*', 'counter-positiveness' (p. 52). Cf. also pp. 3, 39, 42, 99-108, 109-42.
16. For a detailed discussion and explanation for why the number-word *daśa* is symbolically written as a two-digit symbol as 10, see, M.D. Pandit, *Mathematics as Known to the Vedic Saṃhitās*, Sri Satguru Publications, Indian Books Centre, Shakti Nagar, New Delhi, 1993; ch. on 'Zero', pp. 133-77; cf.

also by the same author, *Līlāvatī of Bhāskarācārya*, Part I, published by Sunita Pandit, Behind Ladies' Hostel, University of Poona, Pune, 1992.

17. For greater details, cf. M.D. Pandit, *op. cit.*, pp. 143-46.
18. For details, cf. M.D. Pandit, *op. cit.*, pp. 152-62.

Śūnya in Piṅgala's Chandaḥsūtra

SREERAMULA RAJESWARA SARMA

mayarasatajabhanalagasammitaṃ bhramati vāṅmayaṃ jagati yasya/
sa jayati piṅgalanāgaḥ śivaprasādād viśuddhamatiḥ//

— Halāyudha[1]

The importance of the creation of the zero mark can never be exaggerated. This giving to airy nothing, not merely a local habitation and a name, a picture, a symbol, but also a helpful power, is the characteristic of the Hindu race whence it sprang. It is like coining the Nirvana into dynamos. No single mathematical creation has been more potent for the general on-go of intelligence and power.

— G.B. Halsted[2]

I

The *Chandaḥsūtra* of Piṅgala[3] constitutes an important landmark in the history of the decimal place value system for it mentions, for the first time, zero (*śūnya*) and its symbol. Even though it describes a large number of classical metres in addition to Vedic metres, this text is traditionally considered a Vedāṅga. It is composed in *sūtra* style and contains some 310 *sūtras*, distributed into eight chapters. The descriptive technique employed by Piṅgala has close affinity with that of Pāṇini.[4] Similarity in notation has also been noticed between the *Chandaḥsūtra* and the *Vedāṅgajyotiṣa*. Both these texts employ the first or the final syllable of the names of the *nakṣatras* as their designation.[5] The *Chandaḥsūtra* also introduced a kind of algebraic notation for the classification of prosodic units as *ma-gaṇa, ya-gaṇa* and so on. Significant for the history of numerical notation in India is also the fact that the *Chandaḥsūtra* employs the so-called word-notation or *bhūtasaṃkhyā* quite frequently.

The German indologist Albrecht Weber was the first to make a critical study of the *Chandaḥsūtra* as early as 1863, under the title "Ü ber die Metrik der Inder" in the eighth volume of his journal *Indische Studien*.[6] Since then the text did not receive critical attention, although there is an urgent need of a fresh discussion on the constitution of the text and its chronology.[7]

II

Śūnya is mentioned in the *Chandaḥsūtra* in the context of what are technically known as *pratyayas*.[8] It is, therefore, necessary to give a brief overview of these *pratyayas*. As is well known, the poetic metres in Sanskrit are constituted by the varying arrangements of two kinds of syllables, viz. long syllables (*guru*) and short syllables (*laghu*). *Pratyayas* deal with problems of combinations of these syllabic lengths in a verse-foot (*pāda*). In the last fifteen *sūtras* of the final eighth *adhyāya*, Piṅgala discusses five *pratyayas*, viz. *Prastāra*, *Naṣṭa*, *Uddiṣṭa*, *Saṃkhyā* and *Eka-dvy-ādi-ga-la-kriyā*. In an admirable study, Van Nooten has shown recently how for the first time in history Piṅgala employed here "the binary number as a means for classifying metrical patterns".[9] I may briefly explain each of these *pratyayas*, and against this background dwell at length on the fourth *pratyaya* called *Saṃkhyā* which is relevant to our discussion of *śūnya*.

PRASTĀRA:[10]

It is the arrangement of all possible combinations of *guru* and *laghu* in a given metre. It must be added that Piṅgala deals here only with *samavṛttas*, i.e. metres where all the four feet of the verse have identical patterns. Therefore, he takes into consideration only the verse foot (*pāda*). For example, if the foot of a particular metre contains three syllables, its *Prastāra* will be as shown in following table. Here *g* denotes *guru* and *l* *laghu*.

It will be seen that in the first variation, all the three syllables are *gurus* and that in the last all are *laghus*. How was this table achieved? An examination will show that in the first column, single *guru* (2^0) alternates with single *laghu*; in the second column two *gurus* (2^1) alternate with two *laghus*; and in the third column four *gurus* (2^2) alternate with four *laghus*. This can be continued for any number of syllables in the verse foot. That Piṅgala, however, stops with three syllables has a special reason. These eight variations of three syllables give rise to the so-called *gaṇas* with which the metres are scanned or classified. As I said, Piṅgala designates these eight variations with a kind of algebraic notation: the first variation with all *gurus* is designated *m* or — for the convenience of clear pronunciation — *ma*, or *ma-gaṇa* and so on. In fact, the *Chandaḥsūtra* commences with the definition of these eight *gaṇas* in this very sequence. And now, towards the end, Piṅgala demonstrates how these eight *gaṇas* are achieved.[11]

1.	*g g g*	*m*	*ma*	*ma-gaṇa*
2.	*l g g*	*y*	*ya*	*ya-gaṇa*
3.	*g l g*	*r*	*ra*	*ra-gaṇa*
4.	*l l g*	*s*	*sa*	*sa-gaṇa*
5.	*g g l*	*t*	*ta*	*ta-gaṇa*
6.	*l g l*	*j*	*ja*	*ja-gaṇa*
7.	*g l l*	*bh*	*bha*	*bha-gaṇa*
8.	*l l l*	*n*	*na*	*na-gaṇa*

NAṢṬA:[12]

The second *pratyaya* called *Naṣṭa* is the process of finding out the pattern of a specified variation. Given the serial number of the variation within the *Prastāra*, its pattern has to be found out without actually constructing the *Prastāra*. This is achieved by the continuous halving of the given serial number. Whenever it is divisible by 2, we write down a *laghu* or '*l*'. When it is not divisible, we write down a *guru* or '*g*', add 1 to the number and continue.

For example, we may find out the form of the variation no. 5 in the *Prastāra* for 3 syllables, as we have constructed above. We go on halving the number 5.

5 is odd, write down *g*, add 1 to 5 and halve: $(5 + 1)/2 = 3$
3 is odd, write down *g*, add 1 to 3 and halve: $(3 + 1)/2 = 2$
2 is even, write down *l*, halve: $2/2 = 1$.
Therefore, variation no. 5 has the form *g g l*.

UDDIṢṬA:[13]

In the next *pratyaya* called *Uddiṣṭa*, the pattern is given and its serial number within the *Prastāra* has to be found out. Thus it is just the reverse of *Naṣṭa* and need not detain us.

EKA-DVY-ĀDI-GA-LA-KRIYĀ:[14]

It is the process (*kriyā*) of classifying the variations into those having one *guru*, those having two *gurus* and so on (*eka-dvy-ādi-ga*). For this purpose a pyramid (*meruprastāra*) is constructed in the following manner with $n + 1$ rows, or in the present case with $3 + 1 = 4$ rows.

$$
\begin{array}{ccccccc}
 & & & 1 & & & \\
 & & 1 & & 1 & & \\
 & 1 & & 2 & & 1 & \\
1 & & 3 & & 3 & & 1
\end{array}
$$

The last row provides the answer to the problem. It shows that in the *Prastāra* of 3 syllables there are:

1 form with 3 *gurus* (and 0 *laghu*)
3 forms with 2 *gurus* (and 1 *laghu*)
3 forms with 1 *guru* (and 2 *laghus*)
1 form with 0 *gurus* (and 3 *laghus*).[15]

III

Now we come to the *pratyaya* called *Saṃkhyā* and the mention of *śūnya* therein. This is the method for computing the total number of arrangements of long and short syllables without actually constructing the *Prastāra*. In modern mathematical parlance, this is the number of combinations of two things in 'n' places, repetition being allowed. There are several ways of solving the problem. We have just seen in connection with the *Prastāra* that the number of combinations equals to 2^n where n indicates the number of syllables in the verse foot. Therefore, all one has to do is multiply 2 as many times as there are syllables.[16] Another method is to add up all the numbers in the bottom row of the number pyramid. In the case of three syllables, this will be:

$$1 + 3 + 3 + 1 = 8.$$

But Piṅgala teaches a much more complex method which, as will be shown presently, is of great importance. The method is enunciated in four pithy *sūtras*:

dvir ardhe / rūpe śūnyam / dviḥ śūnye / tāvad ardhe tad guṇitam /[17]

By completing the sentences, these *sūtras* may be rendered into English as follows :

" [First write down the number of syllables in the given metre and go on halving that number. Each time] when [the number is] halved (*ardhe*), [write down in a separate row or column the digit] 2 (*dviḥ*).

"[When you reach an odd number, subtract 1 from it.] Whenever 1 [is subtracted (*rūpe*), write down in a separate column a] zero (*śūnyam*).

"[Continue thus until the process stops. Then where you wrote a] zero (*śūnye*), [multiply by] 2 (*dviḥ*).

"Where [the number was] halved (*tāvad ardhe*), multiply [the result of the second process] by itself (*tad guṇitam*)".

The import of these *sūtras* will be clear from a concrete example. We take the sacred *Gāyatrī* metre which has six syllables in each *pāda*. Required are the number of variations when each one of the six syllables is either a *guru* or a *laghu*. We know already that the answer will be 2^6. But let us see where Piṅgala's method leads to. We proceed as follows:[18]

		A	B	C
1.	Write the number of syllables	6		
2.	Halve 6	3	2	$(2^2.2)^2$
3.	3 cannot be halved; therefore reduce it by 1.	2	0	$2^2.2$
4.	Halve 2.	1	2	2^2
5.	1 cannot be halved; therefore reduce it by 1.	0	0	1.2
		Stop!		Commence!

In column A, the number of syllables is successively halved and, whenever there is an odd number, it is reduced by 1. In column B, on the other hand, we write the two kinds of markers: '2' when halving is possible and '0' when it is not. When halving comes to an end in column A, the process now continues in column C from below, opposite the last marker in column B. Taking unity, double it whenever there is a '0' in column B and square it whenever there is a '2'. At the top of column C, we obtain the result, viz. $(2^2.2)^2$ which is, of course, equal to 2^6.

The process laid down by Piṅgala reduces the number of operations. In the case of *Gāyatrī*, if we try to compute 2^6, we have to multiply 2 by 2 five times, i.e. it involves five operations. Piṅgala's method involves only three: (i) squaring 2; (ii) multiplying by 2; and (iii) squaring again.[19]

Important for the present discussion, however, is the fact that Piṅgala uses the symbols for zero and two as markers for distinguishing between two kinds of operations. The symbol of two marks the place where there is an even number which is divided by 2 and where squaring has to be done later; the symbol of zero marks the stages where there is an odd number and consequently absence of halving and where multiplication by 2 has to be performed. Thus two symbols were used here in a meaningful way. The whole computation can of course be done without any markers at all or with any two arbitrary symbols. However, the fact that Piṅgala uses these two markers shows that at Piṅgala's time there existed a well recognised symbol for mathematical *śūnya*. A symbol presupposes a concept. What kind of mathematical concept lay behind this symbol for *śūnya*? From Piṅgala's use, it may appear that *śūnya* meant here the absence of an operation, akin to the grammarian's *lopa*. But is that all, or does *śūnya* here imply a place value system as well ?[20]

It is useful in this connection to consider the view of Joseph Needham who states that "Place value could and did exist without any symbol for zero. ... But zero symbol as part of the numerical system never existed and could not have come into being without place value".[21]

Therefore, Piṅgala's employment of zero symbol presupposes place value. That it can only be decimal place value needs no emphasis. Vedic literature is replete with decuple terminology such as *eka, daśa, śata* and so forth. To sum up this discussion, Piṅgala could have used any other marker. But that he used a marker called *śūnya* establishes without any doubt that a symbol for zero was quite well known in his times. We do not know what kind of zero symbol it was, but a zero symbol presupposes place value. Therefore, the invention of decimal place value system along with the concept and symbol of zero must antedate considerably Piṅgala's mention of the zero symbol. But when did Piṅgala live?

IV

In an entry on the "Mathematical Aspect of Śūnya" contributed to the second volume of *Kalātattvakośa*, I noted that Piṅgala's time "is variously placed between 400 to 200 B.C." and therefore concluded that "there is enough indirect evidence to say that the decimal place value system with symbols for 1 to 9 and zero developed in India much before the beginning of the Christian era".[22]

This time-frame was contested by a scholar who argued, following Weber, that the *Chandaḥsūtra* was a late work and that, moreover, the eighth chapter where the zero was mentioned was not original.[23]

Although I am reluctant to join the bandwagon of those who are obsessed with establishing the Indian priority in every aspect in the history of ideas, I must say that these two objections are not tenable. It is admittedly difficult to postulate an absolute chronology for any of the ancient Sanskrit texts, but many can be located in relation to other texts. We have noted the close similarity in the method of exposition and notation between the *Chandaḥsūtra* on the one hand and the *Aṣṭādhyāyī* and the *Vedāṅgajyotiṣa* on the other. The affinity with these two texts places the *Chandaḥsūtra* about 400 B.C. But I shall not press this point. Weber's objection to such an early period rests on the fact that the *Chandaḥsūtra* contains several classical metres in addition to Vedic metres. I believe that a re-examination is necessary of the nineteenth century assumption that there existed a clear line of demarcation between the Vedic and classical periods.[24]

As regards the second objection, we are on much firmer ground. *Pratyayas* are discused in the last 15 *sūtras*, i.e. from 8.20 to 8.34, of the *Chandaḥsūtra*. The passage preceding this, viz. *sūtras* 2-19, does not occur in several manuscripts. Therefore, Weber dismissed the entire eighth *adhyāya* as not original. But the absence of *sūtras* 2-19 in some manuscripts does not establish that *sūtras* 20-34 are not genuine. Indeed the whole problematique of *pratyayas* has more to do with Vedic poetry than with the classical poetry. Unlike the classical poetry, Vedic poetry endows the metres with magical significance.[25] Playing with the arragements of metres is a favourite pastime of the Veda and can be understood only in the context of the magical significance of such variations. In his *Rig-Veda Repetitions*, Maurice Bloomfield has shown that variation in metre is a stylistic device in the Veda and that recurrence of otherwise identical *pādas* is accompanied by changes in the metre, which are mostly effected by extensions or abbreviations.[26] Moreover, the various modes of Vedic recitation, such as the *padapāṭha, kramapāṭha, jaṭāpāṭha, ghanapāṭha* deal with arrangements of words in a manner not unrelated to the *pratyayas*. Therefore, it is quite certain that the mathematics of the *pratyayas* developed in the context of Vedic prosody[27] just as the geometry of the *Śulvasūtras* developed in the context of Vedic ritual.

But far more important is the following. The *pratyaya*-section of the eighth chapter is not a loose appendage but is anticipated in the preceding chapters of the *Chandaḥsūtra*. The text commences with the definition of the eight *gaṇas* or triplets, viz. *m, y, r, s, t, j, bh, n,* followed by the definition of *l* and *g*. This sequence of triplets makes sense only in the context of the *Prastāra* of the eighth chapter as I have shown in Section II above. Otherwise, Piṅgala could have enumerated the triplets in any other sequence. Again at the beginning of the fifth chapter, Piṅgala defines the *sama-vṛtta, ardha-sama-vṛtta* and *viṣama-vṛtta* and states that the variations in the second and third categories are obtained by squaring the number of variations in the preceding category (5.1-4). That is to say, since the variations in a *sama-vṛtta* of n syllables are 2^n, the number of variations in an *ardha-sama-vṛtta* will be $2^n \times 2^n = 2^{2n}$; and in a *viṣama-vṛtta* they will be $2^{2n} \times 2^{2n} = 2^{4n}$. This too is not unrelated to the *pratyayas*. Clearly then the eighth chapter containing the combinatorics and the first mention of *śūnya* is not a late interpolation but an integral part of Piṅgala'ṣ *Chandaḥsūtra*.[28]

To conclude, Piṅgala's mention of *śūnya* is a significant event in the history of ideas. It shows that the decimal place value system with the numbers 1 to 9 and zero developed in India before the beginning of the Christian era.

REFERENCES

1. At the commencement of his *Mṛtasañjīvanī* commentary on Piṅgala's *Chandaḥsūtra.*
2. *On the Foundation and Technique of Arithmetic*, Chicago 1912, p. 10.
3. *Piṅgalacchandaḥsūtram*, with the *Mṛtasañjīvanī* commentary by Halāyudha and a further commentary by Jīvānanda Vidyāsāgara Bhaṭṭācārya, Calcutta 1928.
4. B. Van Nooten, "Binary Numbers in Indian Antiquity", *Journal of Indian Philosophy* 21, 31-50, 1993, esp. p. 49, n. 4: "We recognise the use of *adhikāras,*

such as '*chandas*' (2.1), the *anuvṛttis*, the use of *śeṣe* (2.12) to include unnamed contexts, the use of ablative to indicate context-after and the locative for context-before. In brief, enough similarities exist to show that the descriptive techniques used by Piṅgala and the grammarians were similar".

5. David Pingree, *Jyotiḥśāstra: Astral and Mathematical Literature* (J. Gonda, ed, *A History of Indian Literature*, Vol. VI, Fasc. 4), Wiesbaden 1981, p. 10: "It (the *Vedāṅgajyotiṣa*) imitates Piṅgala's Chandaḥsūtra in using the final or first syllables of the names of the nakṣatras as their designations; its period relation is copied in the oldest *Paitāmahasiddhānta*, whose epoch is 11 January 80; and its astronomy reflects that of Mesopotamia in the Achaemenid period. It is likely, therefore, that it was composed not very many years before or after 400 B.C. when the Achaemenids controlled Gandhāra".
6. *Indische Studien: Beiträge für die Kunde des indischen Alterthums*, Band VII, Berlin 1863; reprint: Hildesheim/New York 1973.
7. This is all the more necessary after the publication of the *Chandoviciti*, discovered among the collection of birch bark manuscripts found at Turfan; cf. Schlingloff, Dieter, *Chandoviciti: Texte zur Sanskritmetrik* (Sanskrittexte aus den Turfanfunden, hrsg. Ernst Waldschmidt, Band V), Berlin 1958.
8. On the *pratyayas*, see Ludwig Alsdorf's comprehensive study, "Die Pratyayas. Ein Beitrag zur indischen Mathematik", *Zeitschrift für Indologie und Iranistk*, 9, 97-157, 1933; English tr. by Sreeramula Rajeswara Sarma, "The Pratyayas: Indian Contribution to Combinatorics", *Indian Journal of History of Science*, 26.1, 17-61, 1991. Henceforth references will be made to the English translation.
9. B. Van Nooten, *op. cit.*, 32.
10. *Chandaḥsūtra*, 8.20-23.
11. Piṅgala goes on to add that such triplets can only be eight, neither more nor less. The *sūtra* in question (8.23) reads *vasavas trikāḥ*, i.e. "The triplets (*trikāḥ*) are just eight (*vasavaḥ*) in number". Notice here the use of the word numeral *vasavaḥ* which stands for eight. Piṅgala himself draws our attention to this fact in 1.15: *aṣṭau vasava iti*.
12. *Chandaḥsūtra*, 8.24-25.
13. *Ibid.*, 8.26-27.
14. *Ibid.*, 8.33-34.
15. By tilting the pyramid by about 45°, we get the following figure, which is similar to Pascal's triangle, so named after the French philosopher and mathematician Blaise Pascal, who invented the triangle in 1654 :

1	1	1	1
1	2	3	
1	3		
1			

The four numbers in the hypotenuse of this triangle are, however, the binomial coefficients of $(a + b)^3$ when expanded: $(a + b)^3 = a^3 + 3\,a^2b + 3\,ab^2 + b^3$.

16. Cf. Hemacandra, *Chando'nuśāsana*, 7.11 : *varṇasamadvikahatiḥ*.
17. *Chandaḥsūtra* 8.28-31. The next *sūtra*, 8.32: *dvir dvyūnaṃ tadantānām* is also related to the *Saṃkhyā*, but is not immediately relevant here. It teaches that the sum of the *saṃkhyās* from 1 syllable in a *pāda* to *n* syllables is the *Saṃkhyā* for *n* syllables, multiplied by 2 and reduced by 2, or $2^1 + 2^2 + 2^3 \ldots + 2^n = 2.2^n - 2$.
18. Cf. Datta and Singh, *op. cit.*, i, pp. 75-77; Alsdorf, *op. cit.*, pp. 38-39; see also Gupta, R.C., "An Ancient Method of Piṅgala for Finding a^n", *Bona Mathematica*, 1, 77-80, 1990.
19. Cf. Alsdorf, *op. cit.*, p. 38: "Piṅgala teaches a method which, at first sight, appears to be strange and complicated, but in reality, is very ingenious. It is based, as Weber already explained, 'on a very ingenious way of simplification of the exponent' which results in the reduction of the operations needed for calculating the *saṃkhyā*. For instance, in order to calculate the value of 2^{11}, we have to multiply 2 ten times with 2. But we can go about it in another way. The modern mathematician will factorize 2^{11} thus :

$$2^{11} = 2^{10}.2 = (2^5)^2.2 = (2^4.2)^2.2 = [(2^2)^2.2]^2.2.$$

To calculate the 1st form, we need just to perform thrice squaring and twice multiplication, i.e. five operations instead of 10. The Indian method leads exactly to this kind of simplification of higher powers into lower with an exponent not larger than 2. To know when the squaring is done and when the multiplication with 2, the exponent (in our example 11) is halved continuously: if we reach an odd number, it is reduced by 1. Where an even number is halved, we have to square later and this is marked with an index number '2'. Where 1 is subtracted, we have to perform multiplication later on with 2, and this is marked with the index number 'zero'".
20. Piṅgala employs word numerals rather frequently: in fact, he employs 15 different words with numerical significance in about 100 instances. But in all these places, he uses the words singly. Therefore, his employment of word numerals does not clearly indicate place value.
21. Needham, Joseph and Ling, Wang, *Science and Civilisation in China*, Vol. II, Cambridge 1959, p. 10 n.
22. Sarma, S.R., "Śūnya, Mathematical Aspect" in: Bettina Bäumer (ed.), *Kalātattvakośa: A Lexicon of Fundamental Concepts of the Indian Arts*, vol. 2, New Delhi 1992, pp. 400-11, esp. 403.
23. Bronkhorst, Johannes, "A Note on Zero and the Numerical Place-Value System in Ancient India", *Asiatische Studien/Études Asiatiques*, 48.4, 1039-42,

1994. In support of the view that the *Chandaḥsūtra* is a late work, Bronkhorst cites, on p. 1039, n. 1, half a sentence from Van Nooten, *op. cit.*, p. 33: "... nor is it possible to prove that Piṅgala's work existed before the third century A.D." The other half of the sentence, held back by Bronkhorst, gives a totally different emphasis. Therefore I cite the full sentence: "It is not possible, on objective grounds, to decide whether Piṅgala's treatise preceded or followed Pāṇini, nor is it possible to prove that Piṅgala's work existed before the third century A.D."

24. In the introduction to his edition of the *Chandoviciti*, Schlingloff argues persuasively that the so-called classical metres belong actually to a pre-classical period when variation in metres was the most predominant means of poetic ornamentation as against the *alaṃkāras* of the later times and that classical poetry itself employs only a very small fraction of the large repertoire of metres described in works like the *Chandaḥsūtra.*
25. Cf. Van Nooten, *op. cit.*, p. 31: "The Vedic tradition ascribed a great, almost mystical significance to the metres of the sacrificial chants. Careful studies were made not only on the metres of the chant, but also of its language, prosody, proper place and proper time of recitation. The methodology so developed to study and analyze metres became a respected field of study from a very early time onward. In this tradition the earliest comprehensive treatise on Vedic and Sanskrit metres that has been preserved is the *Chandaḥsūtra* by Piṅgala".
26. Cambridge, Mass., 1916, p. 523. See also, Gonda, Jan, *Vedic Literature (Saṃhitās and Brāhmaṇas)* (A History of Indian Literature, Vol. I, Fasc. 1), Wiesbaden 1975, p. 175 ff.
27. Cf. Bose, D.M., et al (eds.), *A Concise History of Science in India*, New Delhi 1971, pp. 156-57.
28. Van Nooten, whom Professor Bronkhorst cites in order to demolish my chronology, has this to say on the eighth chapter, *op. cit.*, p. 32: "The passage where the binary system developed is in all likelihood part of the original work".

Śūnya in Mathematical System

Zero and the Symbol for Zero in Early Sexagesimal and Decimal Place-Value Systems

DAVID PINGREE

In the early second millennium B.C. Mesopotamian mathematicians used a place-value system of notation with base sixty. Using two wedges impressed on clay — a vertial  for one and a sideways  for ten, they could express by the addition of symbols every number from one to nine and from ten (plus one to nine) to fifty (plus one to nine); after fifty-nine —  — they returned again to , meaning now 1 × 60. Since these numerical forms could be extended infinitely both as integers and as fractions, the symbol  could mean, depending on context, 1, 1 × 60, 1 × 60^2, 1 × 60^3, . . . 1 × 60^n, or 1/60, $1/60^2$, $1/60^3$, . . . $1/60^n$. In the earliest usage of this system, there was no expressed symbol corresponding to an empty sexagesimal place; instead, the Babylonians often avoided confusion by simply leaving an emtpy space between two symbols: e.g., 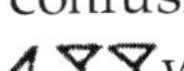 would be read 12, but 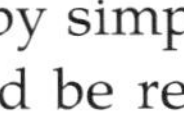 10, 2 = 602. There was no way to write the equivalent of 10, 0, 2 = 36,002; such a number could only be recognized from context.

In the period between the late eighth and the late sixth centuries B.C., some scribes began to use a punctuation mark,  or, , that indicated a stop or separation as the separator of ten or any multiple of ten and of one or any other number up to nine when they were in two sexagesimal places; thus, 10, 2 would be written 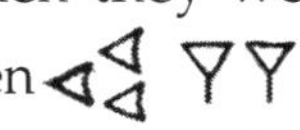 But they also began to use this same symbol to change, for example,  which was either 2, 10 = 130 or 2, 0, 10 = 7210, etc. into the unambiguous , 2, 0, 10. Here was first developed a concrete symbol indicating that there was no number occupying a place in a place-value system (see Neugebauer, 1934-1935, vol. 1, pp. 72-73 [no. 27], and Neugebauer-Sachs, 1945, pp. 34-35 [no. 33]).

This symbol was widely though not universally employed in Babylonian astronomical tables of the last three centuries B.C., in both meanings (see Neugebauer, 1941 and 1955), but the scribes often assumed that the arithmetical structure of the table would guide the reader to the correct interpretation of an ambiguous number. When it was necessary to name an empty sexagesimal place, the Akkadian words used were *nu tuk*, meaning "nothing", the exact equivalent of the Greek οὐδέν which translates it, and of the Latin *nihil*, but not of the Sanskrit *śūnya* and its Arabic derivative, *ṣifr*, both of which signify "empty".

Beginning in the second century B.C. Babylonian astronomical tables and parameters written in this fashion began to be translated into Greek. The Greeks used the letters of their alphabet to indicate numbers, with one letter corresponding to each number from one to nine, one letter for each number from ten in multiples of ten to ninety, and one letter for each number from one hundred in multiplies of a hundred to nine hundred, so that twenty-seven letters expressed all numbers up to 999, after which the sequence repeated itself. Thus, the three letters τξε (300 + 60 + 5) meant 365, for which the Babylonians wrote  (6, 5). The Greeks continued to use their traditional system to write integers in astronomical tables, but for the sexagesimal fractions they wrote the number in each sexagesimal place in their alphabetical notation: thus for $365 + 14/60 + 48/60^2$, which modern historians write in the form 365; 14, 48. Greek papyri of the period immediately preceding and following the beginning of the common era demonstrate that, by using vertical dividing lines such as are also used in Sanskrit manuscripts to separate the sexagesimal places, they removed all ambiguities: τξε|ιδ|μη But they did not leave an empty sexagesimal place blank; rather, they filled it with an adaptation of the Akkadian symbol for zero,  ; this adaptation looks like a circle with a bar over it:  or  (see Jones, forthcoming). This, then, is the origin of the form of the symbol for zero as a circle. It is necessary to note that this form of zero was known, through a translation of Greek astronomical tables into Latin, in Western Europe before the introduction of the Arabic adaptation of the Indian numerals (see Pingree, forthcoming). The Romans, following the Greek practice, wrote the integers as decimal numbers, the fractions as sexagesimals; thus, ρνβ ō ι = CLIIO X = 152; 0, 10.

In India there is evidence in Buddhist and Jaina texts of uncertain date, but near the beginning of the common era, that a decimal place-value system was in use (see Gupta, 1995), but there is no certain evidence that a symbol for zero was in place before the fifth century A.D. However, it is likely that the notational system used in the tables versified in the *Pañcasiddhāntikā* of Varāhamihira — e.g., in the Sine table in IV 6-15, in which the Sines are expressed as units and (sexagesimal) minutes and seconds with zeros in empty places, or the tables of solar and lunar equations in VIII 3 and 6, from the *Romakasiddhānta*, using the same system — reflects the translation of these tables from Greek into Sanskrit in the third or fourth century A.D. (see Neugebauer-Pingree, 1970-1971). This hypothesis of a Greek origin is highly plausible not only because the contents of the tables and their contexts are Greek, but also because the system of notation used by Varāhamihira precisely corresponds to that found in the Greek papyri and medieval manuscripts. All later Sanskrit manuscripts of astronomical texts and tables also universally employ the same system. Since these tables in the *Pañcasiddhāntikā* were based on translations from the Greek, they in all likelihood used the Greek form of the symbol for zero in their sexagesimal fractions.

Exactly when this symbol began to be used in a decimal place-value system is difficult to know because of the lack of documents from the crucial period and the uncertainties regarding the dates and the interpretation of the texts that may be referring to such a symbol. The first indisputable reference, I would suggest, is in the *Paitāmahasiddhānta* of the *Viṣṇudharmottarapurāṇa* (see Dvivedin, 1912), where zero in decimal numbers is indicated not only by two words meaning "empty" — *śūnya* and *kha* (the empty sky) — but also by two words that imply rotundity — *pūrṇa*, "the full (moon)", and *puṣkara*, "a lotus". I have argued elsewhere that the date of the *Paitāmahasiddhānta* is about A.D. 425 (see Pingree-Morrissey, 1989, and Pingree, 1993).

The following facts concerning the steps in the development of a decimal place-value system in India can, I believe, be accepted by all.

1. The numbers in Sanskrit, as in other Indo-European languages, are on base ten.

2. A decimal place-value system in which counting high numbers was achieved by placing counters in a "units pit" indicating units, in a "tens pit" in which they indicated ten each, in a "hundreds pit" in which each indicated a hundred, etc., was being employed in India, especially among Jainas and Buddhists, towards the beginning of the common era (see Gupta, 1992 and 1995).

I would tentatively suggest some further steps.

3. The number of counters in each "pit" would naturally be indicated by a number between one and nine even though each counter in, say, the "hundreds pit" stood for a hundred. This would encourage the writing of those numbers with a sequence of numerals such as 365. However, some "pits" would be empty (*śūnya*). If the numbers were written down as words or as numerals, these empty "pits" would be called by a word meaning "empty", such as *śūnya*, which is itself the symbol for zero.

4. At the beginning of the common era Greeks appeared in India in large numbers, bringing with them, among other things, astronomical tables in which, in the sexagesimal fractions, the "empty" places were occupied by a circular symbol for zero such as are found in the Greek papyri of that period.

5. In translating these tables into Sanskrit, the symbol was taken over as a circle (*pūrṇa*) or a dot (*bindu*).

6. At some time in the period before A.D. 400, someone writing numbers in the decimal place-value system with numerals, used the sexagesimal symbol for zero instead of writing *śūnya*.

At this point India had two place-value systems using the same numerals and the same symbol for zero, while in the West there was only one place-value system, the sexagesimal and it was employed only to represent fractions in astronomical texts and tables. Moreover, the numbers other than zero were written by the Greeks with twenty-seven alphabetical symbols and by the Romans with seven symbols that they came to regard as letters (I V X L C D M). The Arabs in the ninth century learned the Indian numerals with zero and used them in mathematics, but they continued to use their abjad (alphabetical) symbols and sexagesimal fractions with zero in astronomical texts and tables. Similarly when Western Europeans learned the Indo-Arabic numerals with zero in the twelfth century, they replaced the Roman numerals with them in all contexts, but they continued to use sexagesimal fractions in astronomical texts and tables. What they forgot was that those astronomical fractions had been for centuries written with zeros.

REFERENCES

Dvivedin, 1912. V. Dvivedin, *Jyautiṣasidhāntasaṅgraha*, 2 fascs., Benares Sanskrit Series, **39**, Benares 1912, fasc. 2, part 1.

Gupta, 1992, R.C. Gupta, "The First Unenumerable Number in Jaina Mathematics", *Gaṇita Bhāratī*, **14**, 1992, 11-24.

Gupta, 1995. R.C. Gupta, "Who Invented the Zero", *Ganita Bhāratī*, **17**, 1995, 45-61.

Jones, forthcoming. Alexander Jones, *Astronomical Papyri from Oxyrhynchus*. Neugebauer, 1934-1935.

O. Neugebauer, *Mathematische Keilschrift-Texte*, 2 vols., Berlin: Springer, 1934-35.

Neugebauer, 1941. O. Neugebauer, "On a Special Use of the Sign 'Zero' in Cuneiform Astronomical Texts", *Journal of the American Oriental Society*, **61**, 1941, 213-215.

Neugebauer, 1955. O. Neugebauer, *Astronomical Cuneiform Texts*, 3 vols., London: Lund Humphries, 1955.

Neugebauer-Pingree, 1970-71. O. Neugebauer and David Pingree, *The Pañcasiddhāntikā of Varāhamihira*, 2 Vols., Kobenhavn: Det Kongelige Danske Videnskabernes Selskab, 1970-71.

Neugebauer-Sachs, 1945. O. Neugebauer and A. Sachs, *Mathematical Cuneiform Texts*, New Haven: American Oriental Society, 1945.

Pingree, 1993. David Pingree, "Āryabhaṭa, the Paitāmahasiddhānta and Greek Astronomy", *Studies in History of Medicine and Science*, NS **12**, 1993, 69-79.

Pingree-Morrissey, 1989. D. Pingree and P. Morrissey, "On the Identification of the *Yogatārās* of the Indian *Nakṣatras*", *Journal for the History of Astronomy*, **20**, 1989, 99-119.

Pingree, D., *The Preceptum Canonis Ptolomei*, forthcoming.

Chinese "Ling" (0) and Indian "Sunga" (0) Spread to China

WANG YUSHENG

The character *Ling* (Chinese zero) is explained as "occasional drizzle" in the book *Shuo Wen* by Xu Shen (58-147) in the later Han Dynasty. During the Liang and Chen Dynasties of the South Dynasties' period, Gu Yewang (519-581) called it as "scattered showers" in his book *Yu Pian* (543). In the Shong Dynasty, it meant "withered and fallen" in the book *Guang Yun* (751) by Chen Pengnian. As to the *Zi Huei* (1615) by Mei Yiuzhuo of the Ming Dynasty, it meant "odd, with a little extra". Therefore, *Ling* (Chinese zero) has a long history and a lot of meanings in China.

As a mathematical sign, *Ling* (Chinese zero) was expressed by a blank space at first. In the ancient China, Chinese used the decimal system of counting-rods in their numeration, in which the rods could be placed either vertically or horizontally with

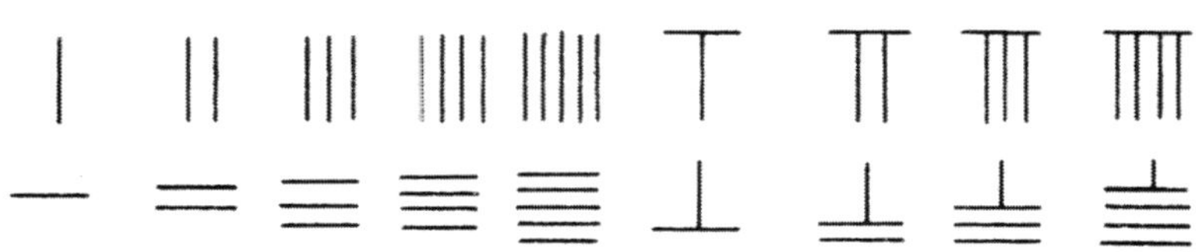

(Vertical)

(horizontal)

equivalent to modern

1 2 3 4 5 6 7 8 9

In fact, the rod digits were always placed alternately vertical or horizontal to avoid confusion in reading: the vertical for units and hundreds, the horizontal for tens and thousands etc.; and the blank space for "zero". Therefore, there was no confusion in reading. For instance,

6 7 8   6 7 8 0   6 7 0 8   6 0 7 8

In the Chinese calendars, such as the Qian Xiang Li (206) by Liu Hong in the later Han Dynasty, the Jing Chu Li (237) by Yang Wei in the Wei Dynasty and the Yuan Jai Li (443) by He Chengtian in the Liu-Song Dynasty, zero was expressed by the blank space besides the character *Chu* (the first). In some cases, other Chinese characters were used to express zero, such as *Duan* (beginning) and *Ben* (origin) in the *Huang Ji Li* (604) by Liu Zhuo in the Sui Dynasty and *Kong* (empty) in the *Wu Yan Li* (624) by Fu Renjun in the Tang Dynasty.

While proof reading old books in ancient China, the unclear or lost characters are used to be symbolized by square frameworks □ which meant the absence of the characters. Later the symbol □ was used to present zero, e.g. Lu Lu Chen Shu (1180):

1 0 6 3 6 8 6 3

The above symbol □ meant zero. In ancient China, the □ was written with writing brush from left to right → and then from top to bottom ↓ them turned from right to left ← and finally from bottom to top ↑□. It becomes zero "0" with quick writing. At least in 13th century, the symbol "0" was used to present zero in mathematics books.

For instance, it was used in Shu Shu Jiu Zhang (1247) by Qin Jiushao of Southern China and Ce Yuan Hai Jin (1248) by Li Yie of northern China. Both of them used the sign 0 in their books. The symbol 0 has been used as zero continuously for about more than one thousand years since then. For example, Xiang Ming Suan Fa (1373), Ce Yuan Suan Shu (1553), Tong Wen Suan Zhi (1613), Shu Li Jing Yun (1723), Bi Wei Suo Yan (1740), Si Yuan Ju Jian Xie Cuo (1829) and Ji Shu Hui Qin (1850) etc. Even recently, O has been included as a Chinese character into dictionaries, such as *Dictionary of Chinsee Language* (1977, 1984) and *Encyclopaedia of Chinese Characters* etc. The Chinese characters are basically block letters. However, the symbol O was accepted rather than the □ as a Chinese character only because it (the Chinese zero) was created independently by China itself.

We know that India already had *śūnya*. Indian numerals spread to China at the beginning of eighth century. Qu Tan Xi Da introduced Indian numerals in a calendar named *Tian Zu*. Jiu Zhi Li Jing included in his book *Kai Yuan Zhan Jing* (729). He said that a dot (.) stood for blank space zero. But it was not used in China. Until thirteenth century, Indian zero sign together with Indian-Arabic numerals, spread to China. The "0" was not used by Chinese as zero at the time anyway. But the symbol "O" was used to present zero. Not until late 19th century or early 20th century, the symbol "0" began to be used as zero instead of "O" in China after many western books had been translated into Chinese.

APPENDIX

The Decimal Place-Value Numeration and the Rod and Bead Arithmetics

Chinese traditional mathematics has since ancient times laid emphasis on practical calculation and made remarkable achievement in this. Worth mentioning are the decimal place-value numeration and the rod and bead arithmetics, which are highly efficient and have done much in promoting mathematics.

Since the existence of written language, Chinese numeration has proved to be positional based on 10. In the oracle-bone inscriptions of the Yin time (c. 14th-11th century B.C.) and bronze inscriptions of the Western Zhou Dynasty (c. 11th century-770 B.C.) all natural numbers within a hundred thousand could be expressed with the aid of a few symbols which stood for one, two, three, four, five, six, seven, eight, nine, ten, a hundred, a thousand and ten thousand respectively plus a few compositions. For instance, the number two thousand six hundred and fifty-six was inscribed as on the bone; and the number six hundred and fifty-nine was on the bronze. The two ancient forms were no doubt decimal place-value forms. They would be quite the same as the modern forms if the few place-value terms were deleted.

The Spring and Autumn (770-476 B.C.) and the Warring States (475-221 B.C.) periods in China witnessed the change from the slave to the feudal society. In agriculture the regular nine-square (in the pattern of the character, or well) farms broke down into small, irregular plots which made land mensuration essential. A more accurate calendar called for precise mathematical calculations. Technical innovation in both agriculture and handicrafts brought about a new upsurge in social productivity which was naturally followed by commodity transactions and the issue of a currency. All these made new ways and means of calculation an acute problem. It was during this period that the counting-rods and rod arithmetic were developed. This is proved by the rod numerals cast on the coins then in circulation; and the appearance of such characters as *suan* and *chou*, or "to calculate" and "the counting-rod" in many literary classics written in that period. (No earlier bone or bronze inscriptions of either are known.) Much cited is the saying of Lao Zi in *Dao De Jing (Canon of the Dao and Its Virtue)* at the end of the Spring and Autumn period: "Good mathematicians can do without the counting-rods".

The best and earliest account of the decimal place-value numeration is found in *Mo Jing (The Book of Master Mo)* written in about 330 B.C.: "One is less than two yet more than five. Explanation is given under 'establishing a position'" (the last 'one' is in a higher-than-unit position); and "there are ones in five and fives in one. The last 'one' is at the next higher place and therefore contains two 'fives'".

According to "Memoir on the Calendar" in *Han Shu (History of the Han Dynasty)*, the counting-rods were round bamboo sticks 0.23 centimetres in diameter and 13.86 centimetres long. Two hundred and seventy-one of these sticks bound together into a hexagonal bundle are conveniently held in the hand. By the 6th century, however, they had become shorter and square or rectangular in cross-section according to *Shu Shu Ji Yi (Memoir on Some Traditions of Mathematical Art)* of the 6th century and also *Sui Shu (History of the Sui Dynasty)*. The changes had been natural since shorter sticks took far smaller space in complicated calculations while square or rectangular shape prevented them from rolling about. Besides bamboo, the sticks could be made of wood, cast iron, jade or ivory. A carrying bag or container was usually attached. In the Tang Dynasty (618-907) all civil and military officers were obliged to carry a bagful of counting-rods with them wherever they went. In recent archaeological excavations a number of ancient counting-rods have been unearthed. More than 30 rods were found in August 1971 in Qianyang County of Shaanxi Province. Dating back to the reign of the Emperor Xuan Di (73-49 B.C.) of the Western Han Dynasty, they are true to the descriptions in *History of the Han Dynasty* except that they are made of bone. The bundle of rods unearthed in early 1975 in No. 168 Han tomb at Fenghuangshan in Jiangling County of Hubel Province are made of bamboo, but a bit longer than the Shaanxi rods. They date from the reign of the Emperor Wen Di (179-157 B.C.).

In rod arithmetic the rods could be placed either upright with equivalent to modern 1, 2, 3, 4, 5, 6, 7, 8, 9 respectively, or horizontally with playing the same roles. In fact. the rod digits were always placed alternately upright or horizontal to avoid confusion in reading: the upright for units and hundreds, the horizontal for tens and thousands etc. For instance, 6,708 would be, the blank space standing for "zero". All arithmetical operations and even algebraic extractions of square and cubic roots could be thus carried out and traced step by step with rods.

The same rod symbol had different absolute values depending on the position in which it occurred. For instance, two thousand two hundred and twenty-two. The whole system was the same as the modern system except for the difference in symbols. In actual calculations the rods were moved about quite like the beads in the later abacus, though with much more difficulty. Systematic treatises on rod arithmetic were *Sun Zi Suan Jing (Master Sun's Mathematical Manual)* written in the 4th century; *Xia Hou Yang Suan Jing (Xiahou Yang's Mathematical Manual)* written in the 5th century, and *Shu Shu Ji Yi* written in the 6th century. In the 1st century, negative numbers appeared in calculations as black negative rods in contrast to the red positive rods. Later rod arithmetic was developed to solving algebraic linear, quadratic, cubic or even higher equations of *n*th order, having well entered the field of algebra. Some equations involved two, three, four or even more unknowns. In short, the counting-rods were important mechanical aids of calculation which enabled ancient Chinese mathematicians to score many great achievements. To find the ratio of the circumference of a circle to its diameter, Zu Chongzhi gave values of π up to six significant figures after the decimal point. He had to calculate for the length of the 12,288 sides of a regular polygon inscribed in a circle and had to extract the square roots of nine-figured numbers for twenty-two times. Zu might not have succeeded had it not been for the aid of decimal positional numeration and rod arithmetic.

Zero in the Mathematical System of India

R.C. GUPTA

INTRODUCTION

Before dealing with the topic or subject of the title, it is obvious that we must be clear of the meaning of the word zero or *śūnya*. Any comprehensive discussion must take into account the various senses in which zero or *śūnya* has been used. A study of the evolution of various concepts behind zero in India will involve investigations in several disciplines which are literary, speculative, as well as scientific. For the convenience of description and discussion, the technology of using zero in India may be treated under the following six broad categories:

(*i*) Literary
(*ii*) Speculative
(*iii*) Referential
(*iv*) Place-value numeration
(*v*) Computable number
(*vi*) Non-scalar or higher mathematical entity.

We shall briefly illustrate these by examples. The dictionary or literal meaning of *śūnya* as 'void' is reflected in the Pali word *sunnāgāra* (*śūnyāgāra* in Sanskrit) or 'empty-room', and metaphorically in Cāṇakya's saying : *avidyaṃ jīvanaṃ śūnyam* or 'life is void without learning'.

Speculative category here includes use in philosophy, mysticism, grammar and linguistics etc. The Buddhist philosophy of *Śūnyavāda* ('Zeroism') springs from the doctrine of non-existence of any Spirit either Supreme or human. The saying *sarvaṃ śūnyam* (all is void) is attributed to the Buddhist Nāgārjuna (about A.D. 150). One of the names of Brahma is *śūnya*. Pāṇini's grammatical or linguistic zero is called *lopa* (*Aṣṭādhyāyī*,

I.1.60) and is used as a maker of an empty non-occupied space.

The referential use of zero is reflected in phrases like 'zero-hour' which is the scheduled time for a specific action or operation. The mean sea-level mark indicates both the zero height and zero depth. *Brahmāṇḍa* may be considered the Potential Zero-point from which the present creation emerged (i.e. Big-bang).

In the place-value systems of numeration (to any base), the *śūnya* as zero denotes the absence of a denominational term. The role of such zero-symbol may be medial or internal (as in 205 or 2005), final or terminal (as in 250 or 2500), or initial (as in 025 or 0025).

The fifth category treats the zero as a number used as computable entity for carrying out calculations and computations. It obeys laws of mathematical operations.

The mathematical or computable zero in the above categories may be confined to the system of real numbers. The next or the sixth category may treat the zero in complex and hyper-complex systems (including quaternions) and also the non-scalar or higher mathematical systems or structures. Examples are zero vector, zero tensor, zero matrix, zero-dimensional commutative rings etc.

EVOLUTION, SPREAD, AND TRIUMPH OF INDIAN DECIMAL PLACE-VALUE NUMERALS

In India, ten had formed the base of counting and numeration since very early time. But for a fully developed decimal place-value system of numerals, we also needed the positional principle as well as the ciphered form of ten numerals (including a symbol for zero). The word *śūnya* (in some sense) is found in the *Atharva-veda* (XIV, 2.19) and the word *kṣudra* found in the same work (*Kāṇḍa* XIX, *sūkta*s 22 and 23) is taken to refer to zero by some scholars. We also know that many words, such as *kham* and *pūrṇa*, which were used later on as technical terms for zero, are found in the Vedas and Vedic literature.

It is claimed that the grammatic system of Pāṇini (about 500 B.C.) had contributed to the evolution of the concept of zero in the mathematical sense, that is, involving, positional analysis, operation of subtraction and process of going from maximum to minimum. Similarly, another Vedāṅga author namely Piṅgala (about 200 B.C. or earlier) is said to have used some symbol for zero. This is inferred from his sūtra *'rūpe śūnyam'* (see his *Chandaḥ-śāstra* VIII, 29) which asks us to place a zero symbol in certain operation while computing 2^n.

It thus appears that the background was ready in India to evolve the decimal place-value notation. And it is believed that the decimal positional system of numerals was available in India before or around the beginning of the present Christian era. This is supported by relevant material found in some ancient Jaina works namely *Anuyogadvāra-sūtra*, *Parikarma* (Kundakunda's lost commentary on the *Ṣaṭkhaṇḍāgama*), *Saṃta-Kamma-Paṃjiya* and *Mahābandha,* all of which are dated before about A.D. 100. Soon the system was used in expressing numbers through the word-numerals as well as the letter-numerals. But due to peculiar Indian social set-up, its knowledge penetrated among the artisans (including the engravers of inscriptions) only slowly.

The Indian decimal place-value system made the writing of even big numbers quite easy and also simplified the arithmetical calculations to a great extent. Foreigners did recognize its merit and the system spread both to the East and to the West. In South East Asia, the Khmere inscription at Sambor (A.D. 683) and the Malay inscription at Palembang (A.D. 684) make use of the system. In China, the Indian system (with *bindu* or thick dot for zero symbol) reached through *Jiuzhili,* the Chinese translation of "*Navagraha Karaṇa*" made by Gotama Siddha (Levensita) during the Thang Dynasty (A.D. 618-907). The first reference in the West to the Indian system appears in a statement of Severus Sebokht (A.D. 662), bishop of the convent at Qenneshre on the banks of Euphrates in Syria. The story of the spread (often slow but steady) and triumph of the Indian numerals all over the civilized world is indeed interesting.[1]

BRAHMAGUPTA'S MATHEMATICS OF ZERO

Simple arithmetical operations involving zero were known and carried out in India earlier than the time of Brahmagupta (A.D. 7th Century). For instance, the *Pañcasiddhāntikā*, IV, 8, of Varāhamihira implies the addition 30 + 0 as *triṃśat yuktāmbara* ('thirty plus zero').[2] But Brahmagupta seems to be the first in the world to give a formal exposition of the *śūnya-gaṇita* (mathematics of zero) in his *Brāhmasphuṭa-siddhānta* (A.D. 628), chapter XVIII (on *kuṭṭaka*). For addition involving positive, negative and zero numbers he states (*BSS*, XVIII, 30):[3]

dhanayor dhanam ṛṇam ṛṇayor dhanarṇayor antaraṃ samaikyaṃ kham /
ṛṇam aikyaṃ ca dhanam ṛṇadhanaśūnyayoḥ śūnyayoḥ śūnyam //30//

'The sum of two positive quantities is positive; of two negative is negative; of a positive and a negative is their (signed numerical) difference; or, if they be equal, zero. The sum of negative and zero is negative; of positive and zero is positive; of two zeros is zero.'

Let the positive and negative quantities (or numbers) be denoted by the letters P and N. Then the verse implies the rules:

$P_1 + P_2 = P_3$; $N_1 + N_2 = N_3$;
$P + N = |P| - |N|$; or $P + N = 0$, if $|P| = |N|$;
$N + 0 = N$; $P + 0 = P$; and $0 + 0 = 0$.

The 4th out of these seven rules may be taken to be Brahmagupta's definition of zero as given by

$$P + (-P) = 0 \ldots (1)$$

The first half of verse no. 32 contains the rules
$N - 0 = N$; $P - 0 = P$; and $0 - 0 = 0$;
and the second half of the next verse the rules:
$0 \times N = 0$; $0 \times P = 0$; and $0 \times 0 = 0$
for subtraction and multiplication respectively.

Then follow two verses (*BSS*, XVIII, 34-35) which contain rules regarding division, square and sqare-root involving positive, negative quantities and zero. Some of these rules are not very or fully clear and some apparently depict mathematical lapse of that remote pre-modern time in which Brahmagupta lived.

One rule in original Sanskrit (see verse No. 34) is

khaṃ khabhaktaṃ kham/ (2)

'Zero divided by zero (becomes) zero.' That is,

$$0/0 = 0 \ldots\ldots (3)$$

which is not proper or sound mathematically. In fact, 0/0 is undefined or indeterminate and its 'value' in the limit is not always 0 as pointed out by Sudhakara Dvivedi in his Sanskrit commentary.[4]

It is interesting to note that Brahmagupta's verse no. 34 is found *verbatim* in the ancient commentary on Śrīdhara's *Pāṭīgaṇita*. And the commentator's explanation of the rule (2) reads[5]

śūnye śūnyena vibhakte phalaṃ śūnyaṃ syāt/

'(When) zero is divided by zero, the result is zero'.

That is, the meaning of (2) is represented by (3).

As for the division of a positive or negative number by zero, Brahmagupta simply calls either of the two ($P/0$ or $N/0$) as '*taccheda*' (that is, a fraction with zero as the denominator) and leaves the matter (verse no. 35). In other words, Brahmagupta defines.

$$x/0 = \textit{taccheda} \; \ldots\ldots \; (4)$$

Some later authors called or denoted $x/0$ as '*khahara*' and tried to understand its nature (see below).

More than twelve centuries after Brahmagupta's correct rule 0 x 0 = 0, an European gave the wrong rule[6]

$$0 \times 0 = 1$$

It is unfortunate that Śrīdhara's larger *Pāṭīgaṇita* (about A.D. 750) is not extant in full and his exposition of zero ('*śūnya-tatva*') is lost. But verse no. 8 in his shorter *Triśatikā* (S. Dvivedi's edition, Benares, 1899) mentions some simple rules about zero including the commutative property:

$0 \times x$ (*khasya guṇane kham* or zero)
$= x \times 0$ (*saṅguṇane khena ca kham*).

MAHĀVĪRA'S TREATMENT OF ZERO

Mahāvīra was a great Digambara Jaina mathematician who flourished in the middle of the ninth century A.D. in South India. His *Gaṇitasāra-saṅgraha* is an excellant work. Its importance is further enhanced due to non-availability of Śrīdhara's larger work. Mahāvīra was the first in the world to state that the square-root of a negative number does not exist among the real numbers. Operations involving zero are mentioned in the following verse from his *Gaṇitasāra-saṅgraha* (I, 49)[7]

tāḍitaḥ khena rāśiḥ khaṃ, so 'vikārī hṛto yutaḥ /
hīno 'pi khavadhādiḥ khaṃ, yoge khaṃ yojyarūpakam //

This has been translated as follows:[8]

'A number multiplied by zero is zero and that number remains unchanged when it is divided by, combined with, or diminished by zero. Multiplication and other operations on zero give rise to zero and in the operation of addition, zero becomes the same as what is added to it.'

That is,

$$x \times 0 = 0$$
$$x/0 = x \quad \ldots . (5)$$
$$x \pm 0 = x$$
$$0 \times x = 0 \text{ etc.}$$
$$0 + x = x.$$

The result (5) is obviously not correct or proper, and shows a mathematical lapse. However, R.N. Mukherjee has interpreted the above verse differently and arrived at the result $0/0 = 0$ instead of (5).[9] Nevertheless, even this interpretion cannot make Mahāvīra free from mathematical slip although Brahmagupta had fallen prey to such a slip earlier.

Regarding the rule (5), we find, the following working line in Mahīdhara's commentary (A.D. 1587) on the *Līlāvatī*[10]

iṣṭam dvayaṃ kalpitam 2, nijārdhena 1 yuktam 3,
tribhir guṇitam 9, khahṛtam 9;...

'Imagine a desired number as 2; combined with its own half, we get 3; multiply by three to get 9; divided by zero, we have 9.'

Mahīdhara's last operation here implies $9/0 = 9$ which is Mahāvīra's (5) with $x = 9$.

ŚRĪPATI'S EXPOSITION REGARDING ZERO

Śrīpati flourished in the middle of the 11th century A.D. and wrote several works. He mentions *Śaka* 961 (= A.D. 1039) as the epoch of his *Dhīkoṭida-karaṇa*. Many operations involving zero are mentioned in verse no. 45 of his *Gaṇitatilaka* as follows[11]

yoge śūnyaṃ bhavati sadṛśaṃ kṣepakasyāvikārī
rāśiḥ śūnyāpagamamilane, śūnyaghāte ca śūnyam /
vyomnā bhakte bhavati gaganaṃ vyomni bhakte ca śūnyaṃ
varge vyomno viyad iti bhaved antarikṣaṃ ghanaś ca //

'On addition of an additive, zero becomes similar (i.e. equal) to the additive. On subtraction or addition of zero, a quantity is unchanged and on multiplication by zero, (a quantity becomes) zero. Zero results on division (of a quantity) by zero as well as on division of zero (by zero).

On squaring of zero, it should be zero. The cube etc. (of zero should also be) zero.'

That is, we have the rules :

$0 + x = x; \quad x \pm 0 = x; \quad x \times 0 = 0$

$x/0 = 0 \quad \ldots . (6)$

$0/0 = 0 \quad \ldots . (7)$

and $0^2 = 0; 0^3 = 0$ etc.

This translation is based on the commentary of Siṃhatilaka Sūri (A.D. 1275) and is supported by the editor of the *Gaṇitatilaka.*[12] In fact the explanatory notes of Siṃhatilaka are quite interesting here. By taking the number (*aṅka*) 5 his illustrations are : $0 + 5 = 5$; $5 \times 0 = 0$. For illustrating (6) he says

khabhaktā daśa kham eva

or (in symbols) $10/0 = 0$

For (7), he states

vyomni śūnye vyomnā bhakte aṅkavat śūnyaṃ bhavati.

'Zero (or *śūnya*) divided by zero becomes zero like an *aṅka* (non-zero number operated similarly).'

That is, the result (7) is similar to the result (6) in which x is not zero.

Relations (6) and (7) are both wrong or objectionable. But while (7) was already given by Brahmagupta, (6) is different from his (4) as well as from Mahāvīra's (5). However, Śrīpati's algebraic chapter (called *Avyaktagaṇita-adhyāya*) XIV, verse no. 6, contains some improved rules as follows (in his astronomical work *Siddhānta-śekhara*)[13]

vikāram āyānti dhanarṇakāni
na śūnyasaṃyoga-viyogatas tu /
śūnyād viśuddhaṃ svam ṛṇaṃ kṣayaṃ svaṃ
vadhādinā khaṃ khaharaṃ vibhaktaṃ //6//

'Positive or negative quantities do not undergo any change through addition or subtraction of zero. On being subtracted from zero, a positive quantity (becomes) negative, a negative quantity (becomes) positive. By multiplication etc. (by zero, a positive or negative quantity becomes) zero. On being divided (by zero, a positive or a negative quantity becomes) *khahara.*'

That is,[14]

$x \pm 0 = x$
$0 - (\pm x) = \mp x$ . . . . (8)
$x \times 0 = 0$
$x/0 =$ *khahara* . . . . (9)

A century later, the famous Bhāskara II used the term *khahara* for $x/0$ similarly as is clear from his statements in his works as follows.[15]

(i) *khabhājito rāśiḥ khaharaḥ syāt* (from *Līlāvatī*, rule 45).
(ii) *khahāro bhavet khena bhaktaś ca rāśiḥ* (from *Bījagaṇita*, section on Zero).

Both these clearly mean that 'a quantity divided by zero becomes *khahara*', that is same as (9).

In the accompanying example in his *Bījagaṇita* (Algebra), Bhāskara II makes it clear that infinity or infinite quantity (*ananta rāśiḥ*) is called *khahara*. He was aware of the fact that the *khahara* remains unchanged by addition or subtraction of any finite number *N*. That is,

khahara $\pm N =$ *khahara* . . . (10)
or symbolically
$(\pm)\ \infty \pm N = (\pm)\ \infty$ . . . . (11)

A PECULIAR RULE FROM BHĀSKARA II

Just after stating $x/0$ as *khahara*, Bhāskara II (about A.D. 1150) continues in his *Līlāvatī* (rules 45-46) as follows[16]

. . . khaguṇaḥ khaṃ, khaguṇaś cintyaś ca śeṣavidhau //45//
śunye guṇake jāte khaṃ, hāraś cet punas tadā rāśiḥ/
avikṛta eva jñeyaḥ, ... //46//

'The product (of a quantity, x) by zero is zero; but the *khaguṇa* (form $x \times 0$) must be retained if further operation impend. Zero having become a multiplier, should zero again become a divisor, the quantity must be understood to be unchanged (*avikṛta*).'
That is, $x \times 0 = 0$ (usual rule), but $(x \times 0)/0 = x$ . . . . . (12)

In this connection the exposition of the commentator Gaṇeśa (son of Keśava) is quite relevant. In his *Buddhivilāsinī* commentary (A.D. 1545) Gaṇeśa states[17] *rāśeḥ śūnye guṇake prāpte tasyānyo vidhiś ced asti tadā khaguṇo rāśiḥ khaṃ syād iti na kāryam/ kintu śūnyaṃ tatpārśve guṇakasthāne sthāpyam iti/ tataḥ śeṣavidhānaṃ kṛte punaḥ khaṃ haraś cet tadā tayoḥ śūnyaguṇakaharayos tulyatvena nāśaḥ kāryaḥ/*

'When a quantity has zero multiplier and other operation is there, then the rule $x \times 0 = 0$ should not be applied. But the zero should be placed by its side as a multiplier. If for the remaining operation, zero is a divisor, then due to equal multiplier and divisor, the zero (in the numerator and denominator) should be cancelled (*nāśaḥ kāryaḥ*)'.

That is, we get (12) by cancelling the zero in the numerator of the lefthand side with the zero of the denominator.

Bhāskara's own example on (12) is to find x from the equation

$$[x \times 0 + \tfrac{1}{2}(x \times 0)] \times 3/0 = 63.$$

This can be written as :

$$(3/2) . (x \times 0) . (3/0) = 63$$

or

$$(9/2) . (x \times 0)/0 = 63.$$

Now by applying (12), we easily get the solution $x = 14$ as given by Bhāskara. Of course, we know that the cancellation rule :

$$(x \times n)/n = x \; . \; . \; . \; . \; (13)$$

is not valid if n is zero (or infinity). But what to say of ancient and medieval writers, even modern scholars have a pitfall in this respect.[18] Actually, the value of the left-hand side of (12) is indeterminate because $x \times 0$ is zero. In fact, once a quantity x becomes zero through multiplication by zero, it can never be reverted to its original form (or value) "just as", in the words of Nārāyaṇa Paṇḍita (about 1350),[19] "a yogin will not revert to rebirth after attaining the blissful state of Brahma".

KṚṢṆA ON VARIOUS RULES ABOUT ZERO

About A.D. 1600 Kṛṣṇa (son of Ballāla) wrote a commentary *Navāṅkura* (also called *Bījapallava* etc.) on the *Bījagaṇita* of Bhāskara II. This commentary gives a good summary of Indian algebra of its time and is written in expository form. Here we present a few things from the 1958 Tanjore edition.[20]

About the nature of zero, Kṛṣṇa says : *khasya dhanarṇatvaṃ nāsti* or

'Zero is neither positive nor negative' (p. 25). The traditional awareness implying the need of commutative property is put by him as *khayogo dvividhaḥ* (p. 24) :

$x + 0$ (*khena yogaḥ*), and $0 + x$ (*khasya yogaḥ*).

Thus $(x + 0)$ or $(a + b)$ is not automatically taken to be same as $(0 + x)$ or $(b + a)$.

About the evolution of the *khahara* $a/0$ (from the fraction a/b) and its nature, Kṛṣṇa remarks (p. 28) (by numerical consideration only) :

yathā yathā bhājakāpacayas tathā tathā labdher upacayaḥ/ tathā sati bhājake paramāpacite labdheḥ paramopacayena bhāvyam/ labdheś ced iyattocyate tarhi paramatvaṃ na syat tato 'py ādhikyasaṃbhavāt/ ato labdher iyattābhāva eva paramatvam/ tad evaṃ upapannaṃ khaharo rāśir ananta iti/

'As much as the divisor (b) is diminished, so much is the quotient (a/b) increased. If the divisor is reduced to the utmost (*parama*), the quotient is increased to the utmost. But, if it can be specified, that the amount of the quotient is so much, it has not been raised to the utmost because a greater than that is possible. Therefore the quotient is great without limit, and the resulting *khahara* quantity is infinite (*ananta*)'.

We may say that he was aware of the limit

$$\lim_{\in \to \infty} \frac{N}{\in} = \infty \text{ (numerically). . . . (14)}$$

Earlier some misunderstanding and confusion prevailed about Bhāskara's statement (10) that the *khahara* remains unchanged by addition or subtraction of finite quantity. For instance, Gaṇeśa (son of Dhuṇḍhirāja), the author of *Gaṇitamañjarī* (circa 1560), combined 11/2 to the *khahara* 4/0 to get[21]

$$(4/0) \pm (11/2) = 8/0 \qquad \text{. . . (15)}$$

by the usual arithmetical process. Thus the *khahara* 4/0 has changed to the *khahara* 8/0 and so Gaṇeśa concluded that there is change (*vikāraḥ*) in the *khahara* contrary to what Bhāskara II believed (apparently the confusion is due to lack of a simple single form of a symbol for *khahara*).

However, Kṛṣṇa refutes the above criticism by saying that although the forms (such as 4/0 and 8/0) are different, their value is same (each being infinity). In other words he showed that (pp. 29-30).[22]

$$a/0 = b/0 \ldots . (16)$$

For this he takes an astronomical problem. We know that the length of the shadow s of a gnomon of length g when the sun's altitude is α is given (in ancient form) by:

$$s = g. (R \cos\alpha)/(R \sin \alpha) \ldots (17)$$

where R is the radius of the circle of reference (chosen arbitrarily). At the sun-rise, the altitude is zero and the shadow will be given by

$$s = g.R/0 \ldots . (18).$$

The usual value of g was 12 *aṅgulas*. For a few values of R, Kṛṣṇa correctly gets the corresponding values of s as follows :

R = 3438	120	100	90
s = 41256/0	1440/0	1200/0	1080/0

Since even with the choice of different radii (*nānātrijyā*), the length of the shadow of a gnomon at any time must be same (*chāya tulyaiva*), he rightly concludes that the above various values of the shadow must be equal. This illustrates the truth of (16). Thus each form of above s represents the same *khahara* (here infinite shadow).

REFERENCES

1. For details, see R.C. Gupta, "Spread and Triumph of Indian Numerals", *Indian Journal of History of Science*, 18 (1983), 23-38.
2. See the *Pañcasiddhāntikā with translation and notes* by T.S. Kuppanna Sastry, edited by K.V. Sarma, PPST Foundation, Madras (now Chennai), 1993, p. 81. For the addition (36 + 0), see p. 337; but the verse XVIII, 35 is controversial.
3. *Brāhmasphuṭa-siddhānta*, edited by R.S. Sharma and his team, Delhi, 1966, vol. IV, p. 1189.
4. *Ibid.*, 1193 where Dvivedi's commentary is given.
5. *The Pāṭīganita of Śrīdharācārya with an Ancient Commentary*, edited by K.S. Shukla, Lucknow, 1959, p. 124.
6. The rule 'nothing multiplied by nothing is one' was given by the Frenchman Antoine L.G. Demonville who wrote about 1831 to 1842; see A. De Morgan, *The Encyclopedia of Eccentrics*, La Salle, 1974 reprint, p. 292.

7. L.C. Jain (editor and Hindi translator), *The Gaṇitasāra-Saṃgraha*, Sholapur, 1963, p. 6.
8. See B. Datta and A.N. Singh, *History of Hindu Mathematics*. Single Volume Edition, Bombay, 1962; Part I, pp. 240-41.
9. R.N. Mukherjee, *Discovery of Zero*, Calcutta, 1991, p. 197.
10. D.V. Apte (editor), *Līlāvatī with the Commentaries of Gaṇeśa and Mahīdhara*, Part I, Poona, 1937, p. 41.
11. H.R. Kapadia (editor), *Gaṇitatilaka with the Commentary of Siṃhatilaka Sūri*, Baroda, 1937, p. 29. Translation by K.N. Sinha in *Gaṇita Bhāratī*, Vol. 4 (1982), 114-133, p. 121.
12. Kapadia, "Introduction" (in reference no. 11 above), p. LVII. Also see pp. 29-30 for Sūri's commentary.
13. B. Misra (editor), *Siddhānta-śekhara* of Śrīpati, Part II, Calcutta, 1947, p. 92.
14. For K.N. Sinha's translation of Śrīpati's algebraic chapter, see *Gaṇita Bhāratī*, Vol. 8 (1986), 27-34, especially, p. 29.
15. *Līlāvatī*, Poona edition (see ref. 10 above), p. 39; and *Bījagaṇita*, edited by J.N. Jha, Banaras, 1949, p. 24.
16. *Līlāvatī* (ref. 10), p. 39. In this connection it has been pointed out that a mistake somewhat serious like Bhāskara's (12) was made by Martin Ohm (1828) who gave

    $(a/0) \times 0 = 0$

    See Datta and Singh, *op. cit.* (ref. 8), p. 246.
17. Gaṇeśa's commentary (ref. 10), p. 40.
18. See Ganitanand, "Some Mathematical Lapses from Āryabhaṭa to Ramanujan", *Gaṇita Bhāratī*, 18 (1996), 31-47; p. 36; and B. Misra (ref. 13), p. 92.
19. In his *Bījagaṇitāvataṃsa*, p. 6. See S.R. Sarma, "*Śūnya*, Mathematical Aspect" in B. Bäumer (editor), *Kalātattvakośa*, Vol. 2, New Delhi, 1992, pp. 410-11 for details.
20. T.V. Radhakrishna Sastrī (Editor), *Bīja Pallavam* of Kṛṣṇa, Tanjore, 1958.
21. See *Siddhānta-śekhara* (ref. 13), p. 18; and *Gaṇita Bhāratī*, Vol. I (1979), p. 24.
22. Also see Datta and Singh, *op. cit.* (ref. 8), pp. 244-245.

Need for Zero in the Numerical System in India

A.K. BAG

The experience of *śūnya* or void on a theoretical or philosophical plane is perhaps common with all the human cultures. I personally feel that it is nothing special of Indian culture. My main objective in this presentation is to assess whether there was any situation where *śūnya* as a part of the numerical system was conceived to tackle problems or simplify a system. George Sarton (1955), the famous historian of science, observes, 'Our numerals and the use of zero were invented by the Hindus and transmitted to us by the Arabs'. Different facets of the problem have already been presented here, but I find no such situation or compulsion needed for such a discovery. On this background I wish to re-examine the whole set of evidences starting from the Vedic period.

VEDIC NUMERALS

The vedic system of numeration was oral. Three-tier system of word numerals were used independently or in combination. The *Taittirīya Saṃhitā* (4.4.11.4), *Maitrāyaṇī Saṃhitā* (2.8.14), *Kaṭhaka Saṃhitā* (39.6), *Yajurveda* (17.2), *Pañcaviṃśa Brāhmaṇa* (17.14.2), *Śāṅkhāyana Śrautrasūtra* (15.11.17) and other Vedic texts give decimal scale along with other numerals. B. Datta[1] has given an excellent summary of the three-tier Vedic word numerals as follows :

1. *eka*(1), *dvi* (2), *tṛ* (3), *catur* (4), *pañca* (5), *ṣaṭ* (6), *sapta* (7), *aṣta* (8), and *nava* (9).
2. *daśa* (10), *viṃśati* (2 x 10), *triṃśat* (3 x 10), *catvāriṃsat* (4 x 10), *pañcāśat* (5 x 10), *ṣaṣṭhi* (6 x 10), *saptati* (7 x 10), *asīti* (8 x 10), and *navati* (9 x 10).

3. *eka* (1), *daśa* (10), *śata* (10^2), *sahasra* (10^3), *ayuta* (10^4), *niyuta* (10^5), *prayuta* (10^6), *arbuda* (10^7), *nyarbuda* (10^8), *samudra* (10^9), *madhya* (10^{10}), *anta* (10^{11}), and *parārdha* (10^{12}).

Features of Vedic Numeration

There is no *śūnya* in the system. This is obvious since the numbers below 100, were obtained by tiers 1 & 2 following additive and sometimes subtractive principles. For numbers above 100, the decimal scale in Tier no. 3 was used in higher to lower order (*sahasra, śata, daśa* and *eka*), along with tiers 1 & 2, for example, *ṣaṣṭhiṃ sahasrāṇi pañca śatāni navatiṃ nava* (= sixty thousand + five hundred + nine ten + nine, i.e., 60599), *sahasrāṇi śata daśa* (= one thousand + one hundred + one ten, i.e. 1110) etc. on additive principle. The names and their order under tiers no. 1 & 2 have been followed by almost all the authorities. Only the number scale under tier 3 has little variation: a term *sayasahassa* (10^5) was introduced by the Jains, and a term *lakṣa* (10^5) used by Buddhists and Āryabhaṭa I were followed by later scholars possibly on the basis of the local uses.

BRĀHMĪ, KHAROṢṬĪ AND OTHER NUMERICAL SYMBOLS (WITHOUT PLACE-VALUE)

The Buddhist text, *Lalitavistara*[2] (100 B.C.) mentions that there were sixty-four *lipis* current durring its time, the first being Brāhmī and Kharoṣṭī. Similar statement is also available in Jain texts. The former was written in left-to-right and the latter from right-to-left order. The Kharoṣṭī alphabets were introduced to north-west of India about 500 B.C., and were used locally in Gāndhāra. Gradually Kharoṣṭī disappeared from India and was replaced by Brāhmī. Majority of the Aśokan edicts are written in Brāhmī from which all other alphabets used in India were evolved. The period from 250 B.C. to A.D. 450 witnessed a large number of distinct symbols in Aśokan, Śaka, Nāgarī, Nāsik, Kṣatrapa, Kuṣāṇa, Gupta, Vallabhī and other inscriptions, a summary of which is given by Smith and Karpinsky from the publications of Bühler, Burnell, Bayley and others (Table 1) :

The table 1 shows that the numbers were expressed by the same additive principle as in the Vedic period and the system is no improvement over the Vedic system. Obviously there was no need for zero in the systems.

Table 1: Numerical Symbols between 250 B.C. and A.D. 450

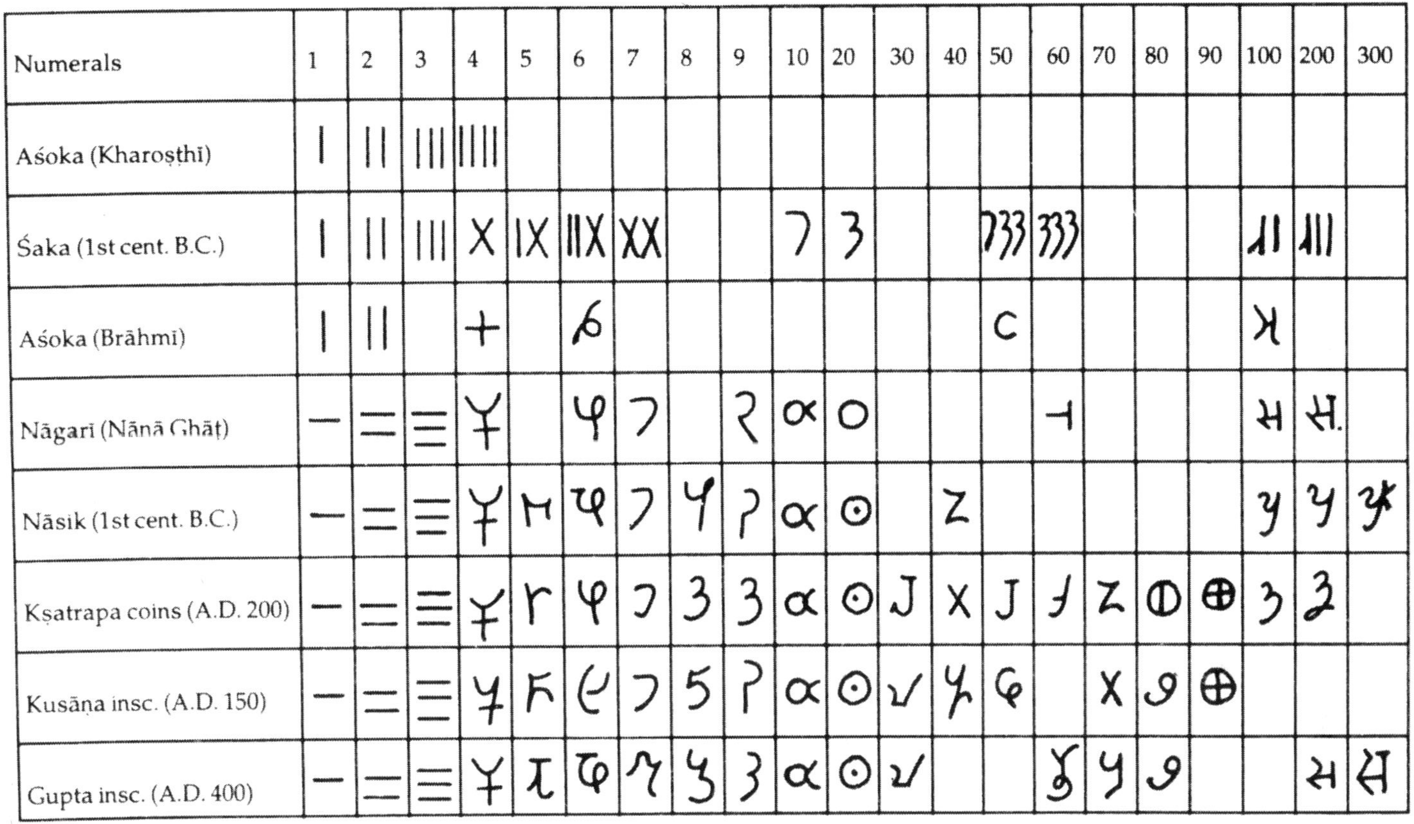

Important Features

The left-to-right and right-to-left traditions made important impact in the later numerical systems in India in which the scales were associated with the places adopting the name place-value. The word-numerals followed the right-to-left order and the symbol-numerals followed the left-to-right order on this palce-value scale, though the number was always counted from higher-to-lower order of the scale creating no problem.

WORD-NUMERALS (WITHOUT PLACE-VALUE)

The Vedic decimal scale mentioned above became a fancy and was widely popular in Buddhist, Jain, Epics and other works. For example, Buddha's training at the hands of his teacher Viśvāmitra runs as follows.[3]

> After me repeat your numeration till we reach the lakh'
> one, two, three, four, to ten and then by tens
> to hundreds, thousands'. After him the child
> named digits, decads, centuries, nor paused,
> the round lakh reached, but softly murmured on,
> then comes the koti, nahut, ninnahut,
> khamba. . . . . ., viskhamha, abab, attata.

These works inform that the students used to recite the decimal scale and followed the same additive principle as was in the Vedas. The word *lopa* was used by Pāṇini and *śūnya* by Piṅgala (A.D. 150). But to what extent these were connected to numerical operation is not clear. But some unique changes also started taking place during this period, specially among the business people.

WORD-NUMERALS (WITH PLACE-VALUE)

The *Anuyogadvāra-sūtra*,[4] first century B.C. Jain work, gives interesting information in its interpretation for revelation of scriptural knowledge. In sūtra 93, it has discussed the properties of ten items serially and one of them is the serial of counting (*gananapubbi*). In sūtra 142 it says that the number of human beings is stated to occupy twenty-nine notational places (*egunatisam-thanai*, in Sanskrit *ekonatriṃśat-sthānāni*).[5] In sūtra 204, it gives the decimal places up to ten for practical purposes serially from the first and from the last. The serial from the first is stated as:

ega(1), *dasa* (10), *saya* (10^2), *sahassa* (10^3), *dasasahassa* (10^4), *sayasahssa* (10^5), *dasasayasahassa* (10^6), *kodi* (10^7), *dasakodi* (10^8), *kodisaya* (10^9), *dasakodisaya* (10^{10}). For serial from the last, it mentioned: *dasakodisaya*(10^{10}) . . . . *ega* (1). This indicated that the decimal scale was decimal places (*sthānāni*) and the order of placement of numeral (*aṅka*) whether dependent on right-to-left or left-to-right order has not been clearly stated anywhere in the work. Further no example is also traceable. This is indeed a unique step. Whatever be the arrangement, this brought a considerable change and simplification even in the system.

The example of the use of decimal place-value is found in the *Yavanajātaka*[6] of Sphujidhvaja (A.D. 269/270). The concluding verse (79.62) has enumerated the date of the text with word numerals in decimal place-value notation as *nārāyaṇa*(1) — *aṅkena* (9) — *indu* (1) mitabdā = 191 śaka year = 191 + 78 = 269 A.D. Actually the order of placement is not clear since first and third positions are denoted by one, which is evident from the following:

Left-to-right order (places):	*eka* (1)	*daśa* (10)	*śata* (10^2)
	nārāyaṇa (1)	*aṅkena* (9)	*indu* (1)=191 śaka
Right-to-left order(places):	*śata* (10^2)	*daśa* (10^2)	*eka* (1)
	indu (1)	*aṅkena* (9)	*nārāyaṇa* (1) = 191 śaka

The left-to-right and right-to-left order continued throughout in Indian tradition. There was hardly any problem with these two systems since the numbers were always counted by higher-to-lower order of the scale following Vedic system. Obviously, the numbers arranged in left-to-right order of the places had to be changed to right-to-left order to get the number in higher-to-lower denominations. It is because of this reason, the experts emphasize the right-to-left order (*aṅkānām vāmato gati*). Mathematical and astronomical texts expressing word numerals are full of such examples.

Varāhamihira (A.D. 505), a descendent of Śakadvīpa brāhmīn who flourished in Ujjain and redacted *Pañcasiddhāntikā*[7] composed during the first three or four centuries of the Christian era, compiled a large number of synonyms of zero, nine numerals from one to nine and a few other selective numerals more than nine, and used them for numbers in left-to-right order, enumerated (Table 2) as follows :

Table 2 : Word-numerals and their Synonyms in Pañcasiddhāntikā

0	*bindu, śūnya, ambara, ākāśa, kha, gagana, viyat, vyoman*
1	*eka, indu, ku, candra, bhū, mahī, rūpa, śaśāṅka, śaśi, śītākara, śītāraśmī, sitāṃśu, himāṃśu*
2	*dvi, akṣi, aśvi, kara, darśa, nayana, pakṣa, yama, yamala*
3	*tri, agni, anala, guṇa, dahana, rāma, vahni, śikhin, hutabhuj, hutāśa, hutāśana, hotṛ*
4	*catur, abdhi, arṇava, kṛta, ghana, jala, jaladhi, caraṇa, yuga, lavaṇoda, veda, samudra*
5	*pañca, akṣa, artha, indriya, bāṇa, viṣaya*
6	*ṣaṭ, ṛtu, rasa*
7	*sapta, adri, aśva, nāga, muni, svara*
8	*aṣṭa, tanu, vasu*
9	*nava, aṅka, randhra*
10	*daśa, āśā, dik*
11	*īśvara, bhava, rudra, śiva, svargeśa*
12	*arka, ina, tīkṣṇāṃśu, dinanātha, divākara, divapa, bhāskara, ravi, maṇḍala*
13	*atijagatī, viśva*
14	*manu*
15	*tithi*
18	*dhṛti*
20	*kṛti, nakha*
21	*mūrcchana*
24	*jina*
26	*utkṛti*
40	*naraka*

The interesting feature is that Varāhamihira (A.D. 505) used both *bindu* and *śūnya* and its various synomyms. The word *aṅka* was used for nine as if there were nine numerals and *śūnya* was not used as a part of the numerals. For expressing numbers, the word-numerals were used with place-value in following order.

Scales :	10^0	10^1	10^2	10^3	10^4
Numbers :	*nava* (9)	— *yama*(2)	— *muni* (7)	= 729	
	kha (0)	— *ṛtu* (6)	— *hutāsa* (3)	= 360	
	kha (0)	— *tithi* (15)		= 150	

This is in left-to-right order, and that the order had to be changed to get the number in higher-to-lower order of the scale. This shows that decimal-place value was known; of course, the word-numerals went up to forty perhaps to maintain metrical rhythm. Since the word *aṅka* denoted the number nine, it justifies that the numerals were in use and there were only nine numerals and the word-numerals *bindu, śūnya, kha* also denoted zero. But there was no precise definition, the main reason being that the problem of numerals were included only in the *Pāṭīgaṇita* (the work dealing with commercial problems and computed on the board). Mahāvīra (A.D. 850) in his *Gaṇitasāra-saṃgraha* gives definitions and synonyms of nine numerals and zero in words and also went beyond ten numerals in decimal place-value system to maintain rhythm. There is also a fact that the numerical symbols were not in common practice. What is the reason? Is it because that the mathematicians were conscious enough for local variation of symbols?

ALPHABETIC NUMERALS AND SYMBOLS

Āryabhaṭa I (b. A.D. 496), the great astronomer-mathematician was aware of this problem. The numerals and their definition are given by Āryabhaṭa I in his *Gaṇitapāda* section (a bit of mathematics) and all later *Pāṭīgaṇita* texts (computation carried out on *pāṭī*) dealing with commercial problems in detail. Āryabhaṭa I was conscious of the local variation of symbols which will be discussed little later. Āryabhaṭa I introduced a completely new system of alphabetic numeration which filled up vacant spaces with zero. In his *Āryabhaṭīya*,[8] he gives the method of writing numbers (verse 1.2) and described the decimal scale as places (verse 2.2), each place being ten times higher than the former. He assigned consonants into two groups, *varga* (*k* to *m* from 1 to 25 respectively) and *avarga* letters (*y* to *h* from 3 to 10 respectively). Nine vowels, *a, i, u, ṛ, l, e, o, ai, au,* will have two sets of values (10^0, 10^1), (10^2, 10^3), (10^4, 10^5), (10^6, 10^7), (10^8, 10^9), (10^{10}, 10^{11}) and so on, first set of values were involved when the vowel is associated with *varga*-letters and the second set when associated with *avarga*-letters. The vacant places were filled up with zero. An example will be of interest in which *A* stands for *avarga* and *V* for *varga*:

Places	A V	A V	A V	A V	A V	A V
Vowels	*e*	*1*	*ṛ*	*u*	*i*	*a*
Values	$10^{11}10^{10}$	$10^{9}10^{8}$	$10^{7}10^{6}$	$10^{5}10^{4}$	$10^{3}10^{2}$	$10^{1}10^{0}$
Numerals :						
(i) *gṛ* = *g* (*varga* let.) with vowel *ṛ*			3	0 0	0 0	0 0
(ii) *khyughṛ* = *kh* (*varga*), *y* (*avarga*) with vowel *u*, and *gh*(*varga*) with vowel *ṛ*			4	3 2	0 0	0 0

Bhāskara I in his commentary[9] on *Āryabhaṭīya* of Āryabhaṭa, gives explanation why Āryabhaṭa adopted alphabetic notation and how zero was written during his time. In his commentary to *Āryabhaṭīya*, verses 2.2., he writes,

kaiṣāṃ sthānānāṃ śaktiḥ, yad ekaṃ rūpaṃ daśa śataṃ sahasraṃ ca bhavati/ satyāṃ caitasyāṃ sthānaśaktau krāyakā viśeṣeṣṭa krayyabhājanāḥ syuḥ, krayyaṃ ca vivakṣāto'lpaṃ bahu ca syāt/ evaṃ ca śati lokavyavahārānyathābhāva prasaṅgaḥ/ naiṣa doṣaḥ/ sthāne vyavasthitāni rūpāṇi daśādīni kṛtāni/ kiṃ tarhi taiḥ? tāni pratipādyante lekhāgamanyāyena/ athavā laghvarthaṃ sthānāni prakramyanta ity uktam asmābhiḥ/ nyāsaś ca sthānānām — 0 0 0 0 0 0 0 0 0 0.

The exact translation runs thus, "What is the power of the places (positions), the symbol one may be ten, hundred, thousand and so on. Truly with these varying values of a (positioned) number, the common consumer may have a possibility of lesser or more quantity of materials purchased. This is so, as the perceptions of local user are different. There is no fault. The figures placed in successive positions take the values ten times of the former. How is that tackled? The alphabetical numerals have been introduced to sort out this problem. After the end of desired positions of alphabets, place 0 0 0 0 0". Āryabhaṭa I was conscious of the local variation of symbols and their problem. In a way, he wanted to have a system which could be used both by astronomers and by ordinary people including traders. By 0 0 0 0, he might have meant that the places were vacant. Āryabhaṭa's method of alphabetical numerals was not very popular either among astronomers or ordinary traders because of the terse methodology but there is no doubt that Āryabhaṭa I discovered the symbol of zero and used it for vacant spaces. The credit of discovering zero goes to Āryabhaṭa I. Bhāskara I was only the exponent since he himself followed the word-numerals. Bhāskara I refers to the works of Maskarī, Pūraṇa, Mudgala, Pūtana and others for their contribution to *Pāṭigaṇita* but their original works are lost.

The *Bakhshālī Manuscript,*[10] another work on *Gaṇita* dealing with commercial problems is found at a village Bakhshali, about eighty kilometres north-east of Peshawar (one of the important trading centres of ancient Gāndhāra district), first used ten symbols for numerals one to nine and zero in decimal place-value system. The influence of Brāhmī or Kharoṣṭī symbols is not unlikely. B.B. Datta who studied the manuscript in great details suggested that the available manuscript is nothing but a commentary and believed it to be a work of A.D. 400 on the basis of internal evidences. Takao Hayashi had examined the whole issue in his Ph.D. thesis on the manuscript and on the basis of *Āryabhaṭīya bhāṣya* of Bhāskara I, Śrīdhara's *Pāṭīgaṇita* (c. A.D. 750) and an anonymous *Tīkā* (edited by K.S. Shukla), not considered before, concluded that the *Bakhshālī Ms.* might have been compiled between 7th and 8th centuries. The symbols used in this are as follows:

A few examples from the manuscript are given below, which show that the numerals were used in left-to-right order.

 (BM, 17 verso) = 330; 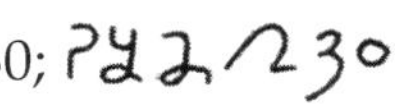 (BM, 56 recto) = 846720

The system also moved to Malaya, Java and other East Indian colonies along with business people which is evidenced from the available inscriptions. The inscription available at Java also contains both symbols and word numerals :

Gurjara Grant inscription (A.D. 595): 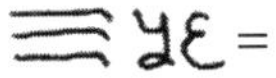 = 346

Khmere inscription at Sambor (A.D. 683): 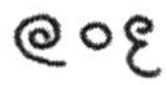 (Śaka) = 605 (Śaka)

Malaya inscription at Palembang (A.D. 684):  (Śaka) = 606 (Śaka)

Malaya inscription at Kotakapur (A.D. 716): 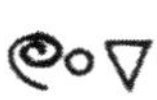 (Śaka) = 608 (Śaka)

Dinaya inscription at Java (A.D. 760):
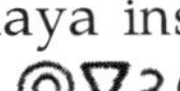 (Śaka) = nayana (2)-vasu(8)-rase(6) Śaka = 682 (Śaka)

CONCLUDING REMARKS

The Vedic system was oral. It could express numbers with the help of a decimal scale from higher to lower order. The Brāhmī and Kharoṣṭī introduced additive system in left-to-right and right-to-left order respectevely. The same left-to-right and right order continued when the decimal scale began to be associated. The two traditions continued with the word numerals in right-to-left order and alphabetic symbol numerals in left-to-right order. There was no apparant contradiction since the numbers were always counted from higher-to-lower order. The word-numeral tradition had conceived zero and had an earlier tradition and started as early as Christian era. The alphabetical and symbol numerals started with Āryabhaṭa I (A.D. 496) when zero was introduced, as reported by Bhāskara I. The full set of nine numerals and zero in symbols are found in the *Bakhshālī Ms.* and other inscriptions between 7th-8th century in connection with commercial problems. Āryabhaṭa I raised the problem of numeration along with business problems and tried to simplify it with a system which needed zero. The nine symbols and zero found in *Bakhshālī Ms.* and commentary on *Pāṭigaṇita*, traced by Shukla were found in current use in north-western region which might have been known to Al-Khwārizmī (c. A.D. 780-850) and Al-Bīrūnī (A.D. 1020). Al-Khwārizmī wrote an arithmetical treatise on Indian system of counting. The original Arabic version is lost, but its Latin translation entitled *De numero Indorum* still survives which has been translated by Yu. Kh. Kopelevich and published in Tashkent in 1964. This has been reported by Rosenfeld.[11] The treatise reads: "Alogorizmi (Al-Khwārizmī) said: when I saw that Indians composed out of IX letters any number due to the position established by them, I desired to discover, God willing, what becomes of those letters to make it easier for the student . . . Thus they created . . . IX letters, figures of them are as follows . . . The beginning of the order is to the right side of the writer, and this will be the first of them consisting of unities. If instead of unity they wrote X, it stood in the second digit, and their figure was that of unity, they needed a figure of tens similar to the figure of unity so that it becomes known that this was X. Thus they put before it one digit and wrote in it a small circle 'o', so that it would indicate that the place of unity is vacant".

What a beautiful description! This reflects the Indian system as reported by Āryabhaṭa I and *Bakhshālī Ms.*

REFERENCES

1. Datta, B.B., "The Vedic Mathematics", *The Cultural Heritage of India,* Ramakrishna Mission Institute of Culture, Vol. 6, reprinted from the Old Series.
2. *Lalitavistara*, eds S. Lefmann and A.S. Halle, London 1902 & 1908, p. 125.
3. 'The Source of Hindu Mathematics,' *Journal of the Royal Asiatic Society*, July, 1910, pp. 749-60, translation by Edwin Arnold, *The Lights of Asia.*
4. *Anuyogadvāra-sūtra*, edited with the commentary of Hemacandra (two parts), Bombay, 1915-16.
5. Sarma, S.R., 'Śūnya/Śūnyatā' *Kalāttvakośa*, vol. II, Delhi, p. 403, 1992.
6. *Yavanajātaka* of Sphujidvaja, edited and translated with notes by David Pingree, Harvard Oriental Series, 2 Vols. 1978.
7. *Pañcasiddhāntikā* of Varāhamihira, with translation and notes by T.S.Kuppanna Sastry and edited with introduction by K.V. Sarma, P.P.S.T. Foundation, Madras, 1993.
8. *Āryabhaṭīya* of Āryabhaṭa, edited and translated into English with notes by Kripa Shankar Shukla, Indian National Science Academy, New Delhi, 1976, pp. 3-5.
9. *Āryabhaṭīya* of Āryabhaṭa with commentary of Bhāskara I and Someśvara, edited by Kripa Shankar Shukla, Indian National Science Academy, New Delhi, 1976, p. 47.
10. The *Bakhshālī Manuscript*, an ancient Indian mathematical treatise, edited and translated by Takao Hayashi, Egbert Forsten, Groningen, 1995; see also Bag, A.K., *Mathematics* in *Ancient and Medieval India*, Chaukhamba Orientalia, Varanasi, 1979, pp. 58-76; Bag, A.K., Mathematical and Astronomical Heritage of India, Project of History of Indian Science, Philosophy and Culture, Occasional Paper no. 14.
11. Rosenfeld, B.A., 'Al-Khwārizmī and Indian Science', vide *Interaction between Indian and Central Asian Science and Technology in Medieval Times*, Vol. I, Indian National Science Academy, New Delhi, 1990, p. 132.

The Mathematical Epistemology of Śūnya

C.K. RAJU

A wrongly perceived *śūnyatā* ruins a person of meagre intelligence. It is like a snake that is wrongly grasped or knowledge that is wrongly cultivated.

Nāgārjuna[1]

Nāgārjuna, Mūlamadhyamakārikā, 24.11[1]

INTRODUCTION

Was Brahmagupta's statement[2] 0/0 = 0 *erroneous*? I will argue that it was not. It has been frequently assumed that (a) *śūnya* means zero, (b) that the concept of zero is uniquely specified by current mathematics, (c) since mathematics concerns universal truths,[3] so that there can be no fundamental differences between mathematics and early Indian mathematics. I will argue that all these assumptions are invalid.

I think it is important to so argue for various reasons. One reason is the recent impressive evidence[4] that the differential calculus was in use in coastal India, at least 200 years before it was first 'discovered' in the West. This makes it important to consider the possibility of *śūnya* as an 'infinitesimal.' Though this possibility was considered by earlier authors, the understanding of 'infinitesimal' in current mathematics has changed after non-standard analysis and computational mathematics.

The Epicurean Ass

Further epistemological complications are illustrated by the story of the Epicurean Ass. Consider the theorem that any two sides of a triangle are together greater than the third.[5] According to Proclus,[6] the Epicureans ridiculed this theorem as evident to any ass. If fodder is placed at an angle to the ass, the ass goes straight to his food, and does not traverse the two sides of a triangle! Proclus' response to this argument may be summarised thus: it is one thing to know *that* a thing is true, and another to know *why* it is true. The ass knows that the theorem is true; the mathematician, who has a proof, knows why it is true.

It has been overlooked that Proclus' response has been applied by western historians of science to press their priority claims with regard to the 'Pythagorean' theorem, as follows. No doubt, the Egyptians, the Indians, the Chinese knew *that* the Pythagorean theorem was true. No doubt, the *Śulva sūtras*, for example, state this proposition;[7] but nowhere in any *Śulva sutra* do we find, a *proof* of the result. But there is at least the rumor[8] that Pythagoras slaughtered an ox when he discovered the proof of this theorem, and there is undoubtedly a proof (or attempted proof)[9] of the theorem in the *Elements*. Hence, while Egyptians, Chinese and Indians can be credited with a knowledge of the Pythagorean theorem, this 'knowledge' was only on par with that of the Epicurean Ass.

Let me hasten to reiterate that my concern here is neither with Pythagoras, nor with the racist and chauvinistic motivations of western historians of science that led to the fabrication of ancient Greece.[10] My concern here is to isolate the central epistemological issue. Is 'proof' essential to mathematics? Is mathematics without 'proof' only proto-mathematics? What exactly constitutes a 'proof'? Is Hilbert's metamathematical notion of 'proof' the only legitimate notion of a valid argument? The Naiyāyikas, the Buddhists, and the Lokāyata did not agree on the notion of 'valid argument' or 'proof'.[11] Neither does the logic used by Nāgārjuna in his *śūnyavāda* agree with the logic of the 'proofs' in the *Elements*, as seen in more detail in § 3. So, is the idea of 'proof' something absolute and 'out there', is it something God-given, or is it a cultural construct, which, in the ultimate analysis, only embodies certain local cultural and aesthetic notions?

Newton presumably thought that 'proof' was essential to mathematics, and that all mathematics must necessarily be modelled on the *Elements*. Hence, he preferred a geometric treatment to a straightforward calculus treatment using infinitesimals. This made his *Principia* so highly obscure, that even 300 years later, people like S.

Chandrasekhar found it challenging enough to write a lengthy exposition of the *Principia*.

Newton's epistemological worries might be combined with the Epicurean-Ass argument to settle the priority claim for calculus in favour of Newton: one might conceivably argue that Mādhava and Nīlakaṇṭha 'only' had some formulae of the calculus, and Jyeṣṭhadeva had 'only' some rationale to offer, while Newton at least attempted 'proof'. The reverse side of the coin is that the calculus fits in only uncertainly with the epistemological scheme for mathematics that Newton had in mind. This only makes it more likely that the calculus was imported from mathematics with a different epistemological tradition, which took western mathematicians two centuries to assimilate.

0/0 in Current Mathematics

With these generalities in the background, let us now turn to the particular statement $0/0 = 0$, from the viewpoint of current mathematics. The reflex reaction of a mathematician is to say that this is wrong. Division by zero is not permitted; it is a common mathematical error, against which a mathematician is trained to guard. As a number, zero is conventionally the additive identity in F, the integral domain (of integers) or field (of rationals or reals) under consideration. Since 0 is the additive identity, $0 + 0 = 0$. From the distributive law in F, it follows that for any x in F, $x. (0 + 0) = x. 0 + x.0 = x.0$, so that $x.0 = 0$, by the cancellation law for addition. Hence, 0 cannot have a multiplicative inverse, so that no meaning may be assigned to $0/0$.

One could try to relax this a bit by using the algebraic properties of the extended reals. To the real numbers R, one attaches two additional symbols ∞ and $-\infty$. Division by zero may now be partially permitted, using the algebaric rules, for example, that $x/0 = \infty$ if $x > 0$. Particularly for the theory of abstract (Lebesque) integration, to permit the neglect of sets of measure zero, we already need[12] at an early stage the convention that $0. \infty = 0$. But we cannot combine the two rules to infer anything about $0/0$ which is still undefined, for we also have the opposite rule $x. \infty = \pm \infty$ if $x \neq 0$. We can, of course, define $0/0$ as we please, for definitions cannot be erroneous. But any convention about $0/0$ would be of doubtful mathematical value because *any* fixed numerical value for the expression $0/0$ would disagree with the limiting value obtained from the differential calculus.

In differential calculus, *l'*Hospital's rule[13] defines the indeterminate form 0/0 in a limit. Suppose f and g are continuously differentiable functions, $f(a) = g(a) = 0$, then

$$\frac{0}{0} \equiv \frac{f(a)}{g(a)} \equiv \lim_{x \to \infty} \frac{f(x)}{g(x)} = \frac{f'(a)}{g'(a)}$$

provided, of course, $g'(a) \neq 0$. On this definition, 0/0 may have any value in the extended real number system.

This suggests that a *useful* or *convenient* definition of 0/0 should regard 0 as infinitesimal. Non-standard analysis[14] motivates the definition of infinitesimals using a field in which the Archimedean property fails. For an ordered field, the Archimedean property says that for any x there exists an integer n such that $x < n$. This property characterizes the real numbers in the sense that the field of real numbers is the largest ordered field in which the Archimedean property holds.[15] Hence, the Archimedean property must fail in any proper ordered-field extension of R. If F is such an ordered field which is a proper extension of R, then in F there exists an x such that $n < x$, for every integer n. Such a number x may be called infinite. Since F is a field and $x \neq 0$, x must have a multiplicative inverse. Its multiplicative inverse is an infinitesimal, for $0 < 1/x < 1/n$ for every integer n. Instead of old-fashioned analytic approach using limits, we could now try to regard 0/0 as the algebraic ratio of two such infinitesimals. In Non-Standard analysis proper, if we divide one infinitesimal by another (we are permitted to do this in an ordered field) we naturally get back *l'*Hospital's rule: in general, the result of the division may have any value ranging from infinitesimal to infinite. That is about as far as current-day formal mathematics can stretch. But there is yet another possibility.

Consider numbers on a computer. The largest (two byte) signed integer one can represent is 32767. The largest (four byte) floating point number one can represent is (of the order of magnitude of) 1E38. Hence, the smallest normal positive floating point number one can represent is 1E-38. Though the floating point numbers are commonly called real numbers (e.g. in FORTRAN), there is *little* in common between the algebraic rules obeyed by floating point numbers and real numbers: e.g. depending upon the conventions used in the design of the hardware and compiler, *even the associative law need not hold for addition or multiplication of floating point numbers.*[16] Let us rename floating point numbers as computational numbers. Then one can say that the arithmetic rules for computational numbers are different from the arithmetic rules for real numbers.

Any arithmetical calculation with computational numbers typically involves discarding or "zeroing" certain non-representable numbers. For the purposes of calculation, both very small and very large numbers are equally non-representable, i.e., *non-representable* refers to *both* infinitesimally small and infinitely large numbers.[17] Furthermore, if x and y are non-representable, then so is x/y, as a rule, though there are exceptions.[18] It is, of course, senseless to have distinct symbols x, y, representing numbers which are treated as non-representable, so the preceding statement could be more consistently and compactly reformulated as $0/0 = 0$, where 0 represents only non-representable number. For any representable number x, it would, of course, continue to be true that $x.0 = 0$, though this equation for computational numbers does not have exactly the meaning assigned to it in the algebraic system of real numbers. (For computational numbers, it incorporates the additional possibility $x. \infty = \pm \infty$, so that *śūnya* here refers also to infinity.) This interpretation of *śūnya* as non-representable is also closer to its meaning in the philosophy of *śūnyavāda*, from which the term was borrowed. Nagarjuna declares: "*śūnya*, non-*śūnya*, 'both,' or 'neither' — these should not be declared. It is expressed only for the purpose of communication."[19]

Moreover, in the algorithmic spirit of Pāṇini, $x.0 = 0$, is a *rule* not a theorem. The difference is that a rule admits exceptions. It cannot be *mechanically* applied; its application may need some intelligence, or at least some contextual understanding.[20] For example, the rule applies to the final step, but intermediate steps are excepted. Thus, Bhāskara II is able to evaluate the value of x (= 44), given[21] that :

$$\frac{x.0 + \dfrac{x.0}{2}}{0} = 63$$

FORMAL VS COMPUTATIONAL MATHEMATICS

Not only do computational numbers differ from real numbers, the epistemology of computational mathematics differs from the epistemology of formal mathematics: in computational numbers 2 + 2 = 4 might mean 2.0 + 2.0 = 4.0000000000000001. But the key difference involves 'proof'. Formal mathematics concerns proofs, while computational mathematics concerns algorithms. Formal mathematicians tend to look with contempt upon computational mathematics: of what use is a computation without a convergence proof? Computational mathematicians tend to reciprocate this contempt. Of what use is it to have an existence proof, if there is no way to calculate the quantity alleged to exist? Moreover, for practical problems involving computational mathematics one can neither wait till a proof is found, nor can one simplify things to the point where there are any existence theorems or convergence proofs in existence.

It is my contention that Indian mathematics was oriented towards calculation rather than proof: it was more computational than formal. I will not argue this contention in detail here, but I will adduce some examples. Brahmagupta understood that the derivative of sine was cosine primarily in the sense that the first differences in a sine table were proportional to cosines. Mādhava, who used the 'Taylor' series to compute a high-precision table of sines and cosines, did not concern himself with a proof of the convergence of the infinite series. These examples can be multiplied.[22]

This is not to say that Indian mathematicians were unconcerned about methodology:[23] in the manner of computational mathematicians they were concerned about rationale and efficient algorithms, but this rationale was not proof. It could not have been 'proof' because the various notions of 'valid argument' in Indian traditions were certainly not identical to the notion of 'valid argument' in the Greek or western traditions.

Proof and Logic

Formal mathematics is deductive while computational mathematics tends to be inductive, or rule-and-exception based. Isn't it true that deduction is certain, whereas rules-and-exceptions and induction are not?

This belief in the certitude of deduction is not consistent with the logic of *śūnyavāda*. This belief in deduction is so unquestioningly accepted in western thought today, that it is instructive first to see it in a historical setting, which shows how the belief has varied with time. A historical horizon for this belief may be found in the commencement of Rational Theology at about the time (*c.* 1250) the West learnt of Aristotle from the Arab rationalists, particularly Ibn Rushd (Averröes). An important concern of rational theologians like Aquinas was to *prove* the *existence* and *uniqueness* of God, to compel the belief of those who did not accept the authority of the scriptures.[24]

Since then, western philosophy has characteristically taken mathematics and logic to represent certitude. After Kant, mathematics and logic are taken to represent *a priori* truth:[25] 2 + 2 = 4 is an *a priori* truth. This trick of treating logic as *a priori* continues in the present-day formalistic tradition of mathematics: all mathematical systems are formal systems based on a two-valued logic.

The most well-known difficulty that this creates is with proofs that use *reductio ad absurdum*. A 'proof' is a finite sequence of statements, each of which is either an axiom, or is derived from two preceding statements by the use of *modus ponens* or instantiation.[26] (Modus ponens is the rule of reasoning which allows B to be inferred from A and $A \Rightarrow B$.) But what does => mean? With 2-valued truth-functional logic, $A \Rightarrow B$, is true, no matter what B is, provided only that A is false. Hence, from contradictory premises, A & $\sim A$, any conclusion B whatsoever may be drawn. Equivalently, there is the *law of the excluded middle* or *law of bivalence: A* or $\sim A$. Within this century, intuitionists have rejected *reductio* proofs as unsatisfactory since non-constructive, particularly in the context of transfinite principles of induction like the Axiom of Choice, Zorn's Lemma, Hausdorff Maximality etc. Most mathematicians have rejected intuitionism, for current mathematics would be impossible without these principles: one cannot, for instance, prove the existence of an uncountable set of measure zero without appealing to these principles. *But this already shows that there is no universal notion of 'proof', even among mathematicians at this time, except through authority and social convention.* Buddhists, however, are well known for their rejection of authority (testimony, Credible Word) in a valid argument.

Against this background, let us now re-examine the basic premise of bivalence. Many will concede that a man might be both good and bad, and this statement might easily be so rendered that it represents a contradiction: does it? For example, let $M(x)$ denote the predicate "x is a man," let $G(x)$ denote "x is good," let "this" be denoted by the ostensive constant value c, so that the statement "this man is good" may be written: $M(c)$ & $G(c)$. Similarly, the statement "this man is bad," might be written: $M(c)$ & $\sim G(c)$. Hence, the statement that "there is man who is both good and bad," may be written: $(\exists x)\ [M(x)\ \&\ G(x)\ \&\ \sim G(x)]$. If this is correct, as it seems to be, one should be able to conclude anything whatsoever from the statement that: "This man is both good and bad."

One can try to escape from this paradox in many ways. For example, one might bring in time: the truth value of a proposition varies with time, and when we say that a man is both good and bad, what we mean is that this man was good at this time in the past and bad at that time.[27] Or one could try and argue that the notions of good and bad are ambiguous. Neither the Jekyll-and-Hyde idea, nor the ambiguity idea really works: Schrödinger's cat is *both* alive and dead at *this* instant.[28] That is about as clear a contradiction as one can get.

Clearly, in resolving this paradox, the need to hang on to 2-valued logic arises only if one wants to preserve the huge structure of current mathematics on which so much effort has been expended. The most natural way to resolve the paradox is to abandon 2-valued logic. There are infinitely many possible logics from which to choose,[29] and one should choose that logic which is most convenient for a variety of purposes, not merely the one that is convenient for preserving past mathematical effort.

The natural approach is exactly what one finds in Indian traditions. B.M. Barua[30] conjectures that prior to the Buddha there was prevalent a theory of four-valued logic. Maurice Walshe refers to these as 'the four "alternatives" of Indian logic: a thing (a) is, (b) is not, (c) both is and is not and (d) neither is nor is not.'[31] This theory of Four Alternatives, which certainly did not apply to all Indian logic,[32] but was frequently used by Nāgārjuna in his famous tetralemma, may be illustrated by an example from the *Brahmajāla Sutta* of the *Dīgha Nikāya*. This Sutta records the Buddha's discourse against various wrong views. The Buddha described four wrong views concerning the nature of the world — whether it is Finite or Infinite — whose adherents claim as follows.[33]

". . . I know that the world is finite and bounded by a circle." This is the first case . . . . ". . . I know that this world is infinite and unbounded". This is the second case. And what is the third way? . . . " . . . I . . . perceive the world as finite up-and down and infinite across. Therefore I know that the world is both finite and infinite." This is the third case. And what is the fourth case? Here a certain ascetic or Brahmin is a logician, a reasoner. Hammering it out by reason, he argues: "This world is neither finite nor infinite. Those who say it is finite are wrong, and so are those who say it is infinite and those who say it is finite and infinite. This world is neither finite nor infinite." This is the fourth case. These are the four ways in which these ascetics and Brahmins are Finitists and Infinitists . . . There is no other way.

Nāgārjuna declares: 'Everything is such, not such, both such and not such and neither such and not such'.[34] It is now easy to see, how miserably *reductio* proofs and bivalence fail in other cultural contexts.

A four-valued logic did not trivialise discourse. A little later in the same *Brahmajāla Sutta* of the *Dīgha Nikāya*, we find the discourse of the Buddha against another of the wrong view labelled as the Wriggling of the Eel.[35]

Because of his dullness and stupidity, when he is questioned he resorts to evasive statements and wriggles like an eel. "If you ask me whether there is another world — if I thought so, I would say there is another world. But I don't say so. And I don't say otherwise. And I don't say it is not, and I don't not say it is not." "Is there no other world? . . ." "Is there both another world and no other world? . . . " "Is there neither another world nor no other world? . . ."

Unlike Ajātasattu's account of Sanjaya Belatthaputta's fivefold negation,[36] as described in the *Samanna Phala Sutta*, we have here clearly a list of *seven* negations: (1) I don't say so, (2) I don't say *otherwise*, (3) I don't say it is not, (4) I don't not say it is not, (5) I don't affirm that there is no other world, (6) I don't say there both is and is not another world, (7) I don't say there is neither another world nor no other world. If we add to this the affirmative proposition of which these are negations, then we obtain the eight possibilities. (It is clearly rather hard to describe so many negations using natural language,[37] though we could think of this as some sort of cyclic negation in an eight-valued logic.)

The relation of time and logic was probably clear to Dignāga who maps a 3 x 3 table onto the Wheel of Reason, which has eight spokes like the eight spokes on the Wheel of Time, with the ninth place being the centre. The second turn of the Wheel, given by the last stanza of his ultra-short treatise,[38] suggests that Dignāga's system of predication is based on three-valued logic of sentences.

If the preference for 2-valued logics is a cultural matter, it would seem that the belief in the universality or absoluteness of deductive proofs is born of mere parochial ignorance of other ideas of what constitutes a valid argument.

CONCLUSIONS

(a) *Śūnya* can be rendered in current mathematics as zero/infinitesimal/ infinity, concepts of which vary in formal algebraic systems, in classical calculus, in non-standard analysis, and in computational mathematics.

(b) *Śūnya* meant "non-representable" both philosophically and mathematically.

(c) Indian mathematics was computational rather than formal. (Though there was a rationale, this rationale was not 'proof' in the supposedly Greek sense.)

(d) Proof, in the current metamathematical sense, is a cultural construct, since the 2-valued logic underlying current mathematics is not universally acceptable. *Śūnyavāda* itself provides an example of a different logic being culturally prevalent.

REFERENCES

1. Tr. adapted from David J. Kalupahana, *Nagarjuna, The Philosophy of the Middle Way* (Sanskrit text of *Mūlamādhyamakakārikā*, with Eng. Tr., Introduction and Annotation), SUNY, New York, 1986, p. 335.
2. Bibhutibhusan Datta and Avadesh Narayan Singh, *History of Hindu Mathematics*, A Source Book, Parts I and II, Asia Publishing House, Bombay, [1935] 1962, p. 245. "Brahmagupta has made the incorrect statement that 0/0 = 0." I do not mean to caricature these authors using a single remark plucked out of context, for they consider in detail the possibility of cipher as infinitesimal.
3. Such beliefs are not exactly dated: e.g. Roger Penrose argues, in *The Emperor's New Mind*, that mathematical truths are 'out there' in a Platonic sense.
4. For an easily accessible account, see George Gheverghese Joseph, *The Crest of the Peacock, Non-European Roots of Mathematics*, Penguin Books, 1991, pp. 270-71, and pp. 286-300. The textual reference to Mādhava's use of the 'Taylor' expansion for the sine and cosine functions, is in Neelkaṇṭha's *Tantrasangraha*, S.K. Pillai (ed.), Trivandrum Sanskrit Series 188, Trivandrum, 1958, p. 19. For a translation of the passage and its rendering in modern notation, see A.K. Bag, "Mādhava's sine and cosine series," *Indian J. History of Sc.* 11(1) 54-7. In addition, one finds explicit expressions for the derivatives of complicated expressions like $\sin^{-1}$ (p sin w) or p sin w/ (1 + p cos w) in *Tantrasangraha* V-53-54, and *Sphuṭanirṇaya* III.19-20, as pointed out by K.V. Sarma (ed. and tr.), *Gaṇitayuktibhāṣā of Jyeṣṭhadeva*, pp. 33-34, unpublished. The significance of 'coastal India' is that particularly Mādhava's accurate sine table could be used in conjunction with Arab-Malayali navigational instruments for accurate departure (*tirfa*) calculations, hence for fixing longitude without a chronometer. See, C.K. Raju,

Kamāl or Rāpalagai, unpublished.

5. This is the 20th proposition of the *Elements*, Book 1. T.L. Heath, *The Thirteen Books of Euclid's Elements*, Dover, New York, 1956, Vol. 1, p. 286.
6. Health, cited above, p. 287.
7. Each *Śulvasūtra* contains a *sūtra* (aphorism, short rule) or proposition to the effect that "the diagonal of a rectangle produces both [areas, squares] which its length and breadth produce separately (Baudhāyana *Śulvasūtra*, i.48; Āpastamba *Śulvasūtra*, i.4; Kātyāyana *Śulvasūtra* ii.11.) together with the proposition that "the diagonal of a square produces an area twice as much [as that produced by its side] (Baudhayana *Sulbasūtra*, i.45; Āpastamba *Śulvasūtra*, i.5; Kātyāyana *Śulvasūtra*, ii.12). The concave of this rule was of practical significance in constructing a right angle.
8. Early Greek historians of geometry, like Eudemus, make no mention of Pythagoras and the first reference to Pythagoras, in the context of geometry, is by Proclus, some thousand years after Pythagoras, who cautiously states, in his commentary on the *Elements*, that rumour credits the theorem to Pythagoras.
9. There are a number of difficulties with the 'proofs' in the *Elements*. The very first proposition requires real numbers (not defined in the *Elements* and certainly non-constructible by ruler and 'collapsible' compasses). Proposition 4 (*Elements*, Book 1) attempts a 'proof' of what is taken as the SAS postulate in current school geometry and in Hilbert's 'synthetic' reconstruction of the *Elements*; use of Birkhoff's 'metric' reconstruction for this proposition would make the proofs of all other propositions prolix and inelegant, since it permits picking and carrying. Proposition 35, on the equality of parallelograms on the same base and between the same parallels, marks a sharp shift from the synthetic to the metric, for it changes the meaning of equality from congruence' to 'equal area', without anywhere defining the notion of area. This undefined notion of area is essentially involved in the proposition which Proclus says 'rumonr' credits to Pythagoras. All this suggests that the *Elements* grew through a gradual process of accretion, with none of its authors sharing exactly the vision of 'proof' used by medieval Rational Theologians and the twentieth century logicians of mathematics.
10. Martin Bernal, *Black Athena. The Afroasiatic Roots of Classical Civilization,* Vol. 1, *The Fabrication of Ancient Greece 1785-1985,* Vintage, London, 1991.
11. S.C. Vidyabhushan, *A History of Indian Logic,* Calcutta, 1921. The Buddhists rejected *testimony* (authority, scripture, Credible Word) as valid *pramāṇa*, while the Lokāyatikas rejected *inference*. (This is not the place to discuss the question whether this rejection of inference was confined to the transcendental sphere,

or extended also to the practical sphere.) The idea that a 'proof' is purely formal is based on the Kantian belief that deduction is a *priori*, while induction is empirical. There is little substance to this belief as argued later on. Hence, I regard *pramāṇa* as comparable to 'proof'.

12. W. Rudin, *Real and Complex Analysis*, McGraw Hill, New York, 1966, pp. 18-19.
13. W. Rudin, *Principles of Mathematical Analysis*, McGraw Hill, New York, 1964, 2nd ed., p. 94.
14. K.D. Stroyan and W.A.J. Luxembourg, *Introduction to Nonstandard Analysis*, Academic Press, New York, 1974.
15. Cohen, *The Structure of the Real Number System*.
16. The limits mentioned in the text can be extended by using more memory to represent numbers; hence real numbers may be represented on a Turing machine which has infinite memory. Further, there are claims that one can use a computer to do arithmetic to arbitrary precision (e.g. William H. Press et al., *Numerical Recipes in C*, Cambridge University Press, 1992, p. 915). However, the term 'arbitrary' must be properly understood, since a necessary condition for the validity of these claims is that one has an infinite amount of *time* (further assuming that time is like the real line, so that an arbitrary amount of it is available in principle) and an unbounded amount of energy at one's disposal, both of which assumptions are naive from the viewpoint of physics (e.g. C.K. Raju, *Time: Towards a Consistent Theory*, Kluwer Academic, Dordrecht, 1994, Chapter 4). While one can no doubt increase the degree of precision and make it very large from a practical point of view, this makes no theoretical difference to our considerations: in philosophical jargon, arithmetic with infinite precision involves a true super task, while performing which a universal Turing machine will never halt.
17. In computer jargon, non-representable refers to both underflow and overflow.
18. Existing C-compilers implement this rule without any exceptions. Thus, consider the following program:
    #include <stdio.h>
    void main (void)
    (float a, b, c; a = 1.0E39; b = –2.0E39; printf ("\n a = %E, b = %E", a, b); getch (); c = a/b; printf ("\n c = %E", c);). The output from this program would be a = + INF, b = –INF, followed by a floating point error.
19. *Mūlamādhyamakakārikā* 22.11; tr. adapted from Kalupahana, cited earlier, p. 307. Here "declared" (*vaktavyam*) is distinct from "spoken" (*kathayate*) or expressed.
20. It may however be mechanically possible (i.e. possible for a universal Turing machine) to recast this contextual understanding into a more elaborate but

comprehensive set of rules as already found in a rudimentary form in object-oriented languages like C^{++}.

21. Datta and Singh, cited above, p. 245. There is a misprint there about the value of x.
22. The *Śulvasūtras* concern metric geometry rather than the synthetic geometry one finds at the beginning of the *Elements*, for *śulva* denotes the rope used for *measurement*. Though there were numerous contacts between Indians and Greeks; and then between Indians and Arabs, who clearly knew about the *Elements*, the *Elements* were not translated into Sanskrit until the eighteenth century. (Presumbably, they were not regarded as worthwhile.)
23. Such a comment has been made for a strange reason pointed out by Srinivas. Under Orientalist pressure, only the 'original' mathematical texts were selectively translated. However, the tradition here required authoritative texts to be very terse, stating the final rules compactly and not going into details of derivation. Only the less authoritative commentaries amplified the derivation of the rule and went into methodological details. M.D. Srinivas, "The Methodology of Indian Mathematics and its Contemporary Relevance," *PPST Bulletin 12* (Sept., 1987) 1-35. However, K.V. Sarma has recently translated Jyeṣṭhadeva's *Yuktibhāṣā*, which concerns rationale as the name suggests.
24. By a strange coincidence, the formal mathematician, too, spends most of his time proving existence and uniqueness! In contrast, Buddhists denied the existence of God or any other Creator, with increasing vehemence from the time of Buddha and *śūnyavāda* is often interpreted as leaning towards universal non-existence, somewhat like Śankara's *Māyāvāda* which it apparently influenced.
25. It is interesting to observe that Kant introduced the notion of *a priori* to respond to Hume whose objections were similar to those of al Ghazāli whom Ibn Rushd tried to answer.
26. For a more formal definition which replaces 'statements' by well-formed formulae etc., see, E. Mendelson, *Introduction to Mathematical Logic*, Van Nostrand, New York, 1964, p. 29.
27. The problem of identity across time is fundamental to Buddhism and time as instant is the key to *Śūnyavāda*: Nāgārjuna declares that *pratitya samutpāda* is *śūnyatā*. Despite the close connection of time and logic, I cannot go into these issues here. Some details may be found in my article, "Time in Medieval India," in: *Science, Philosophy, and Culture*, Part II (D.P. Chattopadhayaya and Ravinder Kumar, eds.), PHISPC, New Delhi, 1997, pp. 253-78.
28. For more details on Schrödinger's cat and its macrophysical empirical

realization in SQUIDS, see my book, cited above. In particular, there is an essential, rather than an accidental, connection between the nature of time and the logic best suited to describe empirical reality.

29. See e.g. N. Rescher, *Many Valued Logic*, McGraw Hill, New York, 1969 ; N. Rescher and A. Urquhart, *Temporal Logic*, Springer, Wien, 1971.
30. B.M. Barua, D. Litt. Thesis, University of London, 1921. In Barua's view this was modified to a fivefold negation by Sanjaya Belatthaputta. B.M. Barua, *A History of Pre-Buddhistic Philosophy.*
31. Maurice Walshe, *The Long Discourses of the Buddha. A Translation of the Dīgha Nikāya,* Wisdom Publications, Boston, 1995, p. 541, footnote 62 to Sutta 1.
32. E.g. the Jaina logic of *syādvāda* (due, I believe, to Bhadrabāhu the junior, roughly a contemporary of Dignāga) had only three basic alternatives: *syāt asti, syāt nāsti, syāt avaktavyaḥ.*The seven categories of *syādvāda* may be expressed as a Cartesian product of these three truth values, as was done by Haldane using the further trick of temporal separation. However, Kothari's belief that the three-valued system of logic might resolve the paradox of Schrodinger's cat was not informed by Reichenbach's similar earlier attempt which was not very successful. J.B.S. Haldane, "The syādvāda system of predication," *Sankhya,* Indian Journal of Statistics, 18 (1957) 195-?. Reproduced in D.P. Chattopadhyaya, *History of Science and Technology in Ancient India*, Firma KLM, Calcutta, 1991, as *Theoretical Fundamentals of Natural Science*, Appendix IV C, pp. 433-40. D.S. Kothari, reproduced as Appendix IV-D, pp. 441-48. More details on Reichenbach's attempt may be found in my book cited earlier.
33. *Dīgha Nikāya*, Tr. Maurice Walshe, cited earlier, pp. 78-79.
34. *Mūlamādhyamakakārikā*, 18.8; Kalupahana, cited earlier, p. 269.
35. *Dīgha Nikāya*, Tr. Maurice Walshe, cited earlier, pp. 80-81.
36. *Evam pi me no, tathā ti pi me no, annathā ti pi me no, iti ti pi me no, notiti pi me no.*
37. For this reason, I am doubtful of the translation of Nāgārjuna's *prasang* into *reductio ad absurdum.* Though it is Nāgārjuna's objective to bring out the absurdity of certain beliefs, *reductio* has a specific meaning today in the context of two-valued logic.
38. Tr. D. Chatterji, "Hetuchakranirnaya," *Indian Historical Quarterly,* **9** (1933), 511-14. Reproduced in full in R.S.Y. Chi, *Buddhist Formal Logic,* The Royal Asiatic Society, London, 1969; reprint, Motilal Banarsidass, Delhi, 1984. See also, Vidyabhushan, cited earlier.

The Importance of the Concept of Zero in Modern Mathematics and Science

J.N. KAPUR

ZERO AS AN INTEGER AND AS A RATIONAL NUMBER

Man first developed natural numbers 1, 2, 3, . . . etc. and soon realised that to every natural number, there is always a successor natural number and he went on to counting numbers up to millions and billions, but in the first stage, he counted only finite numbers. The concept of an infinity of numbers came later. He must have also realised that to every natural number there is a preceding natural number with one exception and that exception was the number 1. It had no preceding natural number. He did not want to have a system which was incomplete and was close-ended at one end. He had to 'invent' a number just preceding the natural number 1. He knew that it was not a counting number and he called it *śūnya*, but as soon as he got *śūnya*, he wanted to have a number preceding *śūnya* and then in the same way invented integers -1, -2, -3, . . . and now he felt a degree of satisfaction with this new system of integers. He could go forward any number of steps and he could go backward any number of steps. He realised for the first time his freedom to create objects without worrying about these objects having a correspondence with objects in his physical experience. It was this freedom to create which was even more responsible for the development of mathematics and civilization than the creation of the number 0 itself.

Another way in which he could have arrived at zero and negative integers was through the operation of subtraction. In the system of natural numbers, he could subtract a smaller number from a larger number, but there was some feeling of incompleteness since he could not subtract a larger number from a smaller number or a number from itself. He invented integers in order to achieve a sense of completeness. He had no other alternative, but to invent negative integers and 0. Again he realised his freedom to create new mathematical objects in this manner.

Thus he developed the system of integers having negative integers, positive integers and 0 and this system had the following properties:

1. For any integer a we have
   $a + 0 = 0 + a = a$
2. For any integer, positive or negative, there exists another integer $- a$ such that
   $a + (- a) = (- a) + a = 0$

This system of integers was mathematically more satisfactory than the system of natural numbers since now he could do the following two operations freely.

1. For every integer he could find the next greater integer, i.e., its successor and he could also find for every integer its predecessor.
2. He could add or subtract every pair of integers and still get an integer.

The desire of getting systems which are complete in themselves for some specified purposes might have motivated the creation of 0 and negative integers. In fact quite often the need for completing the system is so strong that we create new mathematical objects not corresponding, at that point of time, to any physical objects or processes and quite often, it is much later that this correspondence is found.

Cantor said that the "Essence of mathematics lies in its freedom". Man became a mathematician when he first realised this freedom by creating 0 and negative integers.

Mathematicians had no difficulty in extending the operation of multiplication to the system of integers because the product of two integers was always an integer. However there was some difficulty in defining the operation of division, even for the system of natural numbers and even more so in the case of operations with integers, because when we divide one natural number by another natural number or one integer by another integer, the quotient need not be a natural number or an integer. As such, the need arose for defining ratio numbers or what are formally called rational numbers. But even here division by 0 created a problem because if division by 0 is allowed, it could lead to certain contradictions in results in mathematics. The mathematicians could not overcome this difficulty. Of course some persons suggested the idea of an infinity to overcome this difficulty and they suggested that we could call $a/0$ as infinity, but then we would have to call $b/0$ also as infinity and this could imply $a = b$, which is not correct. The mathematicians did not accept this solution, but decided that in order to have a contradiction-free mathematics, division by 0 will not be permitted. They were forced to accept this exception to the general method of division that they developed.

THE ROLE OF ZERO IN THE DEVELOPMENT OF CALCULUS

Of course we could divide 0 by any non-0 number so that $0/a$ gave 0 itself. But then the problem arose as to what is 0/0. Here even the psuedo-solution of calling $a/0$ as infinity was not available. We may also recall at this stage the story about Ramanujan when his teacher told his class that 0 divided by any number gives 0 and he illustrated his statement by saying that if 0 apple is divided among 5 students, each will get 0 apple and if it is divided among 4 students, again each will get 0 apple. Ramanujan at once stood up and he asked his teacher what would happen if 0 apple was divided among 0 students, in that case how many apples will each get. This meant finding 0/0 which the poor teacher had never faced, so that the only alternative for the teacher was to ask Ramanujan to sit down.

However, a similar problem arose with Newton. To just illustrate the problem, consider $2\,x/x$. For all non-0 values of x, it is equal to 2 and therefore it is natural to define $2\,x/x$ as 2 even when x is 0. Similarly consider $-\,x/x$. For all non-0 values of x, it is –l; it is therefore again natural to define it as –1 even for $x = 0$, so as to have no exception. Again consider x^2/x. For all non-0 values of x it is equal to the value of x, as such even for $x = 0$, we should define it as the value of x, i.e. 0. Finally consider $x^{1/2}/x = 1/x^{1/2}$. For $x = 0$ it is not defined. For other values of x, its value is given by the following table :

x	1	1/2	1/3	1/4	1/8	1/16	1/256	1/1024
$1/x^{1/2}$	1	1.41	1.73	2	4	8	16	32

We find that as x becomes smaller and smaller, $1/x^{1/2}$ becomes larger and larger and if we can make x sufficiently small, we can make $1/x^{1/2}$ sufficiently large. Examples like this lead us to the concept of limits and ultimately to the concept of differential calculus which again caused a new revolution in mathematics and science of the same order in magnitude as the earlier revolution created by the invention of 0. Again, it was the freedom of the mathematicians to create new ideas which led Newton and others to make sense out of expressions like 0/0. There was some criticism, but the tremendous success of calculus swept all this criticism aside.

The concept of *śūnya* was very important in the early development of mathematics. However the concept of limit of *śūnya/śūnya* has been equally a great milestone in the development of mathematics and science.

OTHER ZERO ELEMENTS IN MATHEMATICS

So far we have dealt with 0 as an addendum to the system of natural numbers, of 0 as an element of the system of integers. 0 can also be regarded as a rational number which has an infinity of representations namely . . . 0/–3, 0/–2, 0/–1, 0/1, 0/2, 0/3, . . . where a rational number is any number of the form p/q where p and q are integers and q is not 0. This rational number can be regarded as the solution of an equation of the form $qx - p = 0$.

If we solve this first degree equation with integral coefficient and $q \neq 0$, the solution is always a rational number. Now suppose we want to solve a quadratic equation, say $x^2 = 2$ which arises when we want to find the length of the diagonal of a square whose side is unity in length, so that by using Pythagoras theorem $x^2 = 1^2 + 1^2 = 2$ or $x = \sqrt{2}$.

Now it can be shown that $\sqrt{2}$ is not a rational number, i.e. it cannot be expressed in the form p/q where p and q are integers and q is not 0. Since it cannot be expressed as a ratio of 2 integers, it is not a rational number, rather it is an irrational number and if it is expressed as a decimal it will give rise to a non-recurring, non-terminating decimal. Similar irrational numbers arise when we try to find square root or cube root of 3 or 5 or square root of any number which is not a perfect square or cube root of any number which is not a perfect cube. In fact, there are an infinity of numbers which are irrational and the set of irrational numbers together with the set of rational numbers constitute the real number system so that 0 is also a real number. In fact, at that time, the idea of an irrational number appeared to be so strange that Pythagoras and his colleagues who knew that $\sqrt{2}$ was an irrational number kept it a secret because they were afraid of being persecuted, if they talked of any number which was not the ratio of two integers. This indicates the resistance to the creation of new objects in mathematics. Fortunately, such a resistance was much less in India than in may other countries of the world. It took a long time for people to accustom themselves to the idea of irrational numbers and non-recurring, non-terminating decimal numbers.

Introduction of these numbers led to the solution of equations like $2x^2 = 3$, but when mathematicians tried to solve an equation like $x^2 + x + 1 = 0$, they got the solution $x = (-1 \pm \sqrt{-3})/2$. Now they could not digest $\sqrt{-3}$ because there was no known number at that time whose square was negative and they did not have the courage to create new numbers. Even when they created numbers corresponding to square root of negative numbers, they called these as 'imaginary' to placate the opposition. Today these numbers are regarded as real as any other numbers and they have wonderful applications in science and technology. Thus they had to invent numbers like $a + \sqrt{-b}$ where b was positive. They wrote it as $\sqrt{a} + \sqrt{b}\ (\sqrt{-1})$ or as $a + ib$ and called it a complex number with a as real part and b as the imaginary part. Thus 0 also became a complex number $0 = i0$, with 0 both as real part as well as imaginary part.

Thus by gradual extensions, the same 0 became an integer, a rational number, an irrational number, a real number and finally a complex number.

ZERO IN SCIENCE

Mathematicians invented the concept of a vector in order to deal effectively with quantities like velocity, acceleration, force which have both magnitude and direction, but in order to define addition and subtraction of vectors, they had to introduce 0 vector, which has zero magnitude, but is arbitrary in direction. Similarly, when they wanted to discuss addition of matrices, they had to define a zero matrix of every order. In the same way when they had to deal with fluid mechanics, electromagnetism and elasticity, they felt the need for the concepts of stress tensor, strain-tensor and rate-of-strain tensor, each of which had

9 components each in 3 dimensional space. Again they had to introduce a 0 tensor, all of whose components are 0.

Thus science needed the concept of 0 velocity, 0 acceleration, 0 force, 0 mass, 0 volume, 0 temperature, 0 pressure etc. and these were introduced with great care and their introduction was absolutely necessary in order to completely characterise these entities and this would have been impossible without introduction of zero in mathematics by the unknown Indian *ṛṣis*.

In fact these definitions in science were made easier by the earlier definitions of Euclid of a point as an entity with 0 dimension, a line as an entity with 1 dimension and a plane as an entity with 2 dimensions. This was explained by saying that a line consists of an infinity of points each of 0 dimension and due to the infinity of points, it became an object of 1 dimension. Similarly in all areas of science, technology and arts, the concept of 0 was essential to complete the system.

THE ABSTRACT ZERO

All the zeros we have talked about so far refer to relatively concrete situations, but mathematicians also defined an abstract zero by constructing a system of abstract objects related by the same rules of addition and multiplication observed by real number system. They thus got the concept of an abstract field where every abstract field has a 0 with certain properties and this 0 element can be regarded as an abstract 0 of which all the earlier zeroes are relatively concrete representations. In fact all the rules of a field can be followed by a finite number of elements so that we get a finite set of elements obeying all the rules of real number system and having 0 as an element. These finite fields are very useful in communicating messages to satellites and in designing experiments to grow better varieties of wheat. Every field which has an infinity of elements is called a field of characteristic zero and it has been asserted that all important battles of mathematics and science have been fought on fields of characteristic zero.

CONCLUDING REMARKS

Śūnya has been interpreted mathematically, scientifically, philosophically and spiritually in different ways. It has been interpreted as absolute nothingness, as complete absence of light or heat or sound or sensation or complete lack of ego or complete detachment etc.

In mathematics, it has an important role in the place-value system. However its importance rises because without it, we could have only the natural number system. The discovery or creation of 0 opened the floodgates of human creativity in full. Its importance rises because its discovery gave the first sense of freedom to mathematicians to create new objects according to their intuition and the needs of mathematics itself. This freedom has been invaluable to mankind and to science.

Zero is essentially and exclusively a mathematical concept. We cannot have a particle with 0 mass or 0 volume. We cannot have complete vacuum. We cannot have 0 ego. Everything in nature and in man is non-zero. It is only in mathematics that we have the concept of absolute 0 and the dual concept of infinity. In philosophy and religion we also talk of infinity, but here we can not only talk of infinity, but we can have an infinity of infinities and arrange these infinities in a certain hierarchical order. We can talk of an ideal 0 or an ideal infinity only in mathematics. In other disciplines we can have as small objects as we like, but usually there is a limit for the degree of smallness that we can have in nature and similarly we can have a degree of largeness, but there is no infinity in nature. But in mathematics there is no such restriction and we can talk of some variable actually approaching 0 or approaching infinity.

In mathematics we also distinguish between zero and an infinitesimal. While 0 is a fixed number, an infinitesimal is a variable which approaches 0. While we have only one real number 0, we can have an infinity of infinitesimals and we can have a hierarchy of infinitesimals. If x approaches 0, x^2 and x^3 are also infinitesimals, but in some sense x^3 is an infinitesimal of higher order than x^2. In fact it is the relative rates of approaching 0 of infinitesimals which provide the foundation for the very important discipline of calculus.

Moreover the concept of 0 in mathematics is not a static concept. It continuously evolves with time. Mathematics itself grows exponentially and becomes double in a period of 10 years or less. With every growth of mathematics, the concept of 0 also evolves in new directions. It continuously changes. While in other disciplines the concept of *śūnya* can only be refined, in mathematics, it is redefined again and again to include all the new zeros that are evolved with time.

APPENDIX: THE IMPORTANCE OF ZERO

1. The importance of the creation of zero mark can never be exaggerated. This which gives to airy nothing, not merely a local habitation and a name, a picture, a symbol, but helpful power, is characteristic of the Hindu race from which it sprang. It is like coining the Nirvāṇa into dynamos. No single mathematical creation has been more potent for the general on-go of intelligence and power.

   — Halstead

2. It is India which gave us the ingenious method of expressing all numbers by means of ten symbols, each symbol receieving a value of position as well as an absolute value, a profound and important idea which appears so simple to us now that we ignore its true merit. But its very simplicity, the great ease which it lent to all the computation, puts our arithmetic in the first rank of useful inventions and we appreciate the grandeur of this achievement the more, when we remember that it escaped the genius of Archimedes and Apollonius, two of the greatest men produced by antiquity.

   — Laplace

3. The problem of numeration was formally solved by the Hindus at some controversial date before 800 A.D. The introduction of zero as a symbol denoting the absence of units of certain powers of ten in a number represented by the Hindu numerals has been rated as one of the greatest practical inventions of all times.

   — E.T. Bell

REFERENCES

1. J.N. Kapur, A Brief History of Hindu Mathematics and Astronomy in *Fascinating World of Jawaharlal Nehru University and Mathematical Sciences*, Vol. VII, pages 60-78, Mathametical Science Trust Society, New Delhi.
2. D.S. Hooda and J.N. Kapur, *Āryabhaṭa*, New Age Int. Pub., 1996.

The Concept of Śūnya and Its Ramifications in Mathematics and Computer Science

M.R. ADHIKARI

INTRODUCTION

For the genesis of the concept of zero in mathematics, one has to go back in the early centuries of the Christian era and to have a recourse to the then Indian culture of education. The basic concept of *śūnya* (zero) originated in Indian Vedic literature and supported by the philosophical views[1] fluxed in the veins of Indian culture, formed the spine on which the first mathematical definition of *śūnya* leading to axiomatic statements of logistics thereof was built by Brahmagupta (A.D. 628).[2] This paper will show how the concept of *śūnya* founded by Brahmagupta, maintained by Śrīdharācārya (A.D. 750) and Śrīpati (A.D. 1039) and renovated by Bhāskarācārya II (A.D. 1114-1200) branched out into ramifications in the number systems, the algebraic system, topology and computer science.

RAMIFICATION OF THE DEFINITION AND LOGISTIC OF ŚŪNYA IN MODERN MATHEMATICS

(A) Number System

We are familiar with positive integers as natural numbers from our childhood through the process of counting and start mathematics with them in an informal way by learning to count followed by learning addition and multiplication, the latter being a repeated process of addition.

The set N of natural numbers, as we know today, can be determined by Peano's axioms (A.D. 1889) and the set Z of integers is logically obtained from the set N by introducing the concept of integers: the integer zero (0) and negative integer $(-n)$ of a natural number n, each of which is determined uniquely in Z. Both these concepts, namely the integer zero (0) and the negative integer $(-n)$ of a natural number n along with some operations on them were stated by Brahmagupta (*c* A.D. 628) in the following line :

dhanarṇayor antaraṃ samaikyakhaṁ[3]

i.e., the sum of one positive and one negative each of equal magnitude, is zero. In the language of algebra, if a is a positive number (a natural number, as we have said) then $a + (-a) = 0$.

Undoubtedly, Brahmagupta's statement echoed in the larger field of modern number system and the extension of N to Z can be described as a ramification of Brahmagupta's mathematical concept of zero.

(B) Algebraic System

Before examining the case of ramification of ancient Indian logistics of *śūnya* in the field of modern algebra we recall some relevant basic concepts.

(i) *An algebraic system* is a non-empty set S with one or more operations on S.

(ii) *A semigroup* is an ordered pair $(S, *)$, where S is a non-empty set and $*$ is an associative binary operation on S.
The concept of the integer 0 (zero) in Z is extended to the zero-element in an algebraic system[4] $(S, +)$ by the defining property: $0 + a = a + 0 = a$ for all a in S, where + denotes a binary operation on a non-empty set S.

(iii) A semi-group $(S, +)$ with zero element is said to be a *monoid*.

The concept of an abstract additive group $(G, +)$, is defined with a view to generalising the number-system $(Z, +)$ endowed with the usual addition. The zero element 0 and the negative of an element in an additive group $(G, +)$ is determined uniquely by the property :

$$0 + a = a + 0 = a \text{ for every } a \text{ in } G \qquad \ldots (1)$$

and the negative $-a$ of a in G is determined uniquely in G by the property:

$(-a) + a = a + (-a) = 0$ . . . (2)

Moreover $0 + (-a) = 0 - a = -a$ for all a in G . . . (3)
and $-(-a) = a$ for all a in G . . . (4)

Here the property (1) reminds us Śrīdharācārya's statement:

kṣepasamaṃ khaṃ yoge rāśir avikṛtaḥ khayojanāpagame[5] [in the process of addition (of a number) to zero, if zero is added (to a number) or subtracted (from a number) the number does not change, i.e., for a number a, $a + 0 = 0 + a = a$].

The property (2) was stated by Brahmagupta. The negative of a number was defined by Śrīpati[6] and also by Bhāskarācārya II in a slightly different language but more explicitly and there are clear manifestations of the properties (3) and (4) in an algebraic system : $(Z, +)$

(i) *śunyād viśuddhaṃ svam ṛṇam kṣayaṃ svam'*. . .

(A number, when subtracted from zero gives the negative of itself and the negative of the negative of a number is the number itself)

i.e., for a number a, $0 - a = -a$
and $-(-a) = a$

(ii) *khayoge viyoge dhanarṇaṃ tathaiva cyutaṃ śūnyatas tad viparyāsam eti* (Bhāskara II) (When zero is [added to or] subtracted from a positive or a negative number, the number remains as it is; a number subtracted from zero undergoes a change in sign)

i.e., for a number a, $a + 0 = a$, $(-a) + 0 = -a$,
$0 - a = -a$, $0 - (-a) = a$.

In modern mathematics, the phrase 'zero element' has a sense broader than that used by ancient Indian mathematicians. In fact, the zero element denoted by the symbol 0 may have varied meanings (different from the customary meaning in India) depending on the binary operations to be defined on S. The following examples will clarify this point :

Example (1) : Consider the monoids $(N, +)$ and $(N, .)$ where N is the set of all non-negative integers and $a + b$ is defined by :

$a + b$ = (l.c.m of a, b where both a and b are positive integers)
= 0, otherwise

and $a.b.$ = the usual product of non-negative integers a and b.

Then, $a + 1 = 1 + a = a$ for all a in N, shows that the integer 1 is the additive zero element of (N, +) but the integer 0 (zero) is the multiplicative zero element of $(N, .)$.

Example (2) : Consider the monoid $[P(S), U]$ of power set of the non-empty set S. Taking the set-theoretic union of sets as the binary operation on $P[S]$, the empty set $\emptyset$ is the zero-element of $[P(S), U]$ as

$\emptyset \; U \; A = A \; U \; \emptyset \; = A$ for every A in $P(S)$

(C) Topology[8]

For a pointed topological space X, if C denotes the constant loop $C : I -$ X, carrying all of $I = [0, 1]$ to the point x_0 of X, the homotopy class $\dot{C}$ of C is the zero element of the fundamental group $\pi_1 \; (X, x_0)$.

(D) Computer Science[9]

Our main interest in the computer science is on languages. The free monoid M^* generated by a finite set M is the set of all finite products $uvw, \ldots.,$ where $u, v, w, \ldots,$ are elements of M. The elements of M are called '*letters*' and those of M^* are called words, M being itself an '*alphabet*'. The word '*free*' means that two words are considered to be the same only when they consist of the letters of the same order. The role of zero element is played by the empty word e with the property that $ew = we = w$ for every word w of M^*. The subset $M^* \;\; U\{e\}$ determines the language.

CONCLUSION

The discovery of *śūnya* (zero) in numerals is a magnificent Indian contribution. The first mathematical definition of *śūnya* was given by Brahmagupta in the seventh century. Moreover the concept of the fundamental operations with *śūnya* in mathematics in a logical and systematic way was developed in India by Brahmagupta, *Śrīpati*, Bhāskarācārya II and others during the period A.D. 628 to A.D. 1200. The concept of *śūnya* developed in India in the environment of her cultural tradition led to the modern concept of zero element in mathematics and mathematical sciences. Indian concept of *śūnya* has wide ramifications in the growth of science, technology and civilization.

ACKNOWLEDGEMENT

The author expresses his sincere thanks to Indian National Science Academy of New Delhi, for extending financial support to the project under which this work was carried out.

REFERENCES

1. Mukhopadhyay, A., Impact of Indian Philosophy on Bhāskarīya, Concept of Zero. Read in the Seminar on Concept of *Śūnya*, New Delhi, on Feb. 12-14, 1997.
2. Bag, A.K., *Mathematics in Ancient and Medieval India*, Chowkhambha Orientalia, Varanasi, 1979, page 19.
3. Sharma, Ramaswarup, *Brāhmasphuṭa Siddhānta*, Indian Institute of Astronomical and Sanskrit Research, New Delhi, 1966, pp. 243-44.
4. Adhikari, M.R., *Groups, Rings and Modules with Application,* Universities Press, 1999, pp. 62.
5. Shukla, K.S., *Pāṭīgaṇita of Śridharācārya*, p. 14
6. Misra, Babuaji, *Siddhānta Śekhara* by Śrīpati, Calcutta University Publication, 1947, p. 92.
7. Dwivedi, Sudhākara, *Bījagaṇita* by Bhāskara II, Krishnadas Academy, Varanasi, 1983, pp. 13, 14.
8. Munkres, J.R., *Topology — A first course*, Prentice Hall of India Pvt. Ltd., 1978.
9. Adhikari, M.R., *Groups, Rings and Modules with Application,* Universities Press (India) Ltd., 1999, pp. 208.

Impact of Indian Philosophy on Bhāskarīya Concept of Zero

A. MUKHOPADHYAYA

INTRODUCTION

In *Ṛgveda*[5] (about 3000 BC)[6] the two words (a) *śūnam* (S-1) and (b) *śūne* (S-2)* are available each having a direct bearing on the present form of the word *śūnya* (zero) and the word '*śūnam*' though finally means a 'swollen state' or 'hollowness', has been taken to imply lack, want or adherence. In *Atharva veda*,[8] *śūnya* (zero) been used (S-3) to denote emptiness, absence etc. *Atharva veda* praising the king Parīkṣit and his son Jañmejaya[9], the two descendents of the Pāṇḍavas in *Mahābhārata*, may be taken for a work either contemporary of or subsequent to *Mahābhārata* which refers to a period not later than that of Haryaṅka-Śaiśunāga kings[10] (600-500 B.C.) of Magadha. The term *śūnya* (zero) in the sense of *abhāva* (i.e., lack, nullity, absence etc.) is also found to occur in *Rāmāyaṇa* (*śūnyaṃ ratham*, verse 2.52.41); in *Śatapatha Brāhmaṇa* a later class of Vedic work (verse 2.3.1.9), in *Manu Siddhānta* (*naikaḥ supyācchūnyagehe* . . . verse 57, *adhyāya* 4). The meaning of the term has been mentioned explicitly in *Nāmaliṅgānuśāsana* (*Amarakoṣa*), a Sanskrit lexicon written by Amarasiṃha of the 5th century A.D. (S-4) thus :
i.e., *śūnya* (zero) (is) also (called) *vaśikam* (void or empty), *tuccha* (negligibly small, trifling), *riktaka* (emptied or devoid of).

---

* Verses S-1, S-2, are all given in the Appendix.

RECOGNITION OF ZERO AS A NUMBER AND ITS SYMBOLIC REPRESENTATION

Piṅgala's *Chandaḥ Sūtra*[12] (prior to 200 B.C.)[13] particularly the pair of verses 28 (*dviraddhe*) (S-5) and 29 (*rūpe śūnyam*) (Ṡ-6), manifests that *śūnya* (zero) in the sense of *abhāva* was regarded as a number at least in the days of Piṅgala. That symbolic notations of zero as well as of numbers were in practice in those days will be evident from the lines :

(i) *dvisaṃkhyāṃ bhūmau prastārayet* (the number 'two' should be represented on the floor) and

(ii) *tasminn apanite śūnyaṃ labhyate tat pūrvalabdhāyā dvisaṃkhyāyā adhastāt sthāpayet* (subtracting that when 'zero' is obtained, that zero should be placed below the number 'two').

The original *Bakhshālī Manuscript* (written in 3rd century A.D.[14]) employs *śūnya* (zero) symbolically to mean void or empty in decimal arithmetical[15] calculations. But the exact type of symbol used so far in India is not known. According to Datta,[15] the earliest symbol for *śūnya* (zero) was a dot and the small circular symbol was introduced in Indian mathematics before the eighth century A.D. Puliśa (of pre-Śaka-period[16]) in his *Puliśa Siddhānta*[17] regarded *śūnya* (zero) as a numeral and incidentally mentioned addition of zero to a number (S-7).

In the early medieval period (400-1200 A.D.) when arithmetic and algebra along with geometry were grouped together under the name of *Gaṇita*, different from *Jyotiṣa*[18], demanding intensive study of the system of number, the complete decimal arithmatic in Bhāskara-I)'s[19] (574 A.D.[20]) commentary of the *Āryabhaṭīya* formed the background on which Brahmagupta (628 A.D.) endeavoured to put a rigorous treatment of the mathematics of zero.

MATHEMATICAL DEFINITION OF ZERO

Brahmagupta[21] was the first to advance a mathematical definition of zero (S-8). The sum of two positive numbers is positive, the sum of two negatives is negative, the sum of one positive and one negative is the difference and the sum of two equals (opposite in signs) is zero, i.e., for a number a, a-a = 0, where by zero (*śūnya* or *kha*) he meant void (*śūnyam ākāśam*[22]) or *riktaka*, in conformity with the philosophy of *abhāva* a traditional concept of zero even in the early seventeenth century A.D.

(Nṛsimha in 1621 A.D.) though by this time Bhāskarācārya (I) put forward his superfine concept of zero as an indefinitely small quantity (tending towards nullity as the limiting value), as if, taking into consideration the meaning *tuccha* of *Amarakoṣa*. This article will throw light on the philosophy of *abhāva* after Kaṇāda in his *Vaiśeṣika Darśana* and the Nāgārjuna-philosophy of *śūnyavāda* — its impact, if any, on Bhāskara (I) enabling him to get out of the limitations of the former philosophy.

Kaṇāda's Clarification of Abhāva

In the chapter *abhāva parīkṣā*[23] of *Vaiśeṣika Darśana*, Kaṇāda (or Ulūka), an Indian philosopher in the age of *Mahābhārata* (not later than the fifth century B.C.) defined *abhāva* in the following way :

It (*abhāva*) is (i) the state of never-occurrence of a thing, even in traces (*yat kiñcin na kadācit bhavati*) or (i) the state in which a trace of a thing seldom occurs somewhere; but not everywhere, not even always in any form whatsoever (*kiñcic ca kadācid bhavad api kenacid rūpeṇa kutracid bhavati, na sarvadā, na rūpantareṇa na khalv api sarvatra)*. The first category of *abhāva* has been termed non-materialistic (*avastubhūta*) while the later class has been called materialistic *abhāva* (*vastubhūta*) (S-9).

According to the commentator Chandrakānta, (a) the state of non-occurance (*yan na bhavati*) or of absence (*yan na vidyate*) of a thing or (b) the state of abandoning a thing which was existent before (*prāg utpatteḥ parataś cātmaprahāṇanna bhavati*) is a materialistic *abhāva* (S-10).

Śūnyavāda in Nāgārjuna Philosophy

According to Nāgārjuna's philosophy[24], the primary meaning of *śūnyatā* is the sense of devoidness, which is a direct reference to the truth of things mundane (worldly) and ultimate, but it refers also to the method by which *śūnyatā* as truth is brought to light by rejecting the imagination of the ultimacy and absolutenese of particular entities and of specific concepts and conceptual systems. It is not a rejection of the world of becoming and of the meaningfulness of life but the very mundane existence as a course of conditioned becoming as well as a way through which the values of life become possible of realisation.[25] According to Kumārajīva, *śūnyatā* does not mean non-existence or nothingness or complete extinction. The ultimate nature is devoid of origin and is impermanent. This ultimately true nature of things, i.e., complete *śūnyatā* (as Kumārajīva emphasized) is the same as *nirvāṇa* and *nirvāṇa* 'compasses end and beginning,' 'It is not attained by attaining,' 'The way of attaining it by not attaining'[26] is the main theme. As pointed out by Radhakrishnan,[27] Nāgārjuna's *nirvāṇa* is not something to be attained. It is the ultimate goal towards which all beings move seeking fulfilment.

Concept of Ākāśa (or Kha) as Synonym of Zero: Views from the Two Philosophical Angles :

The mathematical definition of *śūnya* (zero) (*dhanarṇayor antaraṃ samaikyakhaṃ*, i.e., the sum of two equal quantities differing in signs, is zero), as was introduced by Brahmagupta, follows the idea of *abhāva* in the sense of abandoning a number which was existent before (*prāg utpatteḥ parataś cātmaprahāṇanna bhavati*). The concept of *śūnya* was clarified as *śūnyamākāśam* where the words *śūnyam* (zero) and *ākāśam* (the sky) were taken for synonymous. According to Kaṇāda as Chandrakānta points out, *ākāśa*[28] makes room for entrance or exit (*ākāśo hy avakāśam dadāti tato niṣkrāmati praviśati ca*), but it is not a place (or an object) where the two associated operations of entering into or going out are possible (*karma khalv etat niṣkramaṇaṃ praveśanaś caikasmin dravye samavaiti, yan niṣkrāmati praviśati ca na tv etad ākāśe 'pi samavaititi*). It is ever-existing and so it has no beginning (*vyatirekābhāvāt pūrvabhāvaniyamana*), it is also omnipresent (*sarvagato'pi*) and singular (*ekatve cākāśasya*). It is a cause for making room for space having no material confinement (*mūrta saṃyoga bhāvas tatra hetur evam ākāśo 'py avakāśanimittaṃ hetur bhavati*). It is a continuous idea, a name which cannot be split due to variance in its appearance (*khalv ekam ākāśaṃ bhāva vad upādhi bhedābhidyate*).

According to Nāgārjuna philosophy, *ākāśa* is a name, a permanent principle of accommodation without resistance, it is formless and without any specific character. It is of the same nature as *nirvāṇa*,[29] the ultimate goal to which all beings tend for fulfilment.

Division by Zero, Brahmagupta's Key to Compass a New Aspect of Zero :

It is not clear whether Brahmagupta's concept of *ākāśa* had any bearing on the above two philosophical doctrines, but it is true that the problem of division of a number by zero, which perplexed Indian mathematicians before him, was given due attention in his *Brāhmasphuṭa Siddhānta* and he introduced the term *taccheda* to mean the form $a/0$ where '*a*' denotes a positive or a negative number (*khoddhṛtam ṛṇadhanam vā tacchedaṃ*, i.e., zero upon a positive or a negative is *tacchedam*). Clearly he could not come to a definite conclusion about the nature of the form $a/0$ (*kha* and *ākāśa* being synonymous) when '*a*' is non-zero. Also $0/0$ was improperly interpreted as zero (*khaṃ khabhaktaṃ khaṃ*, p. 245). These limitations were due to the inadequacy of the concept of *śūnya* as void though this concept posed no problem before Brahmagupta for defining all other operations dealing with zero exhaustively. More than five hundred years after Brahmagupta, Bhāskarācārya II (A.D. 1114) put forward a new way of interpretation of 'division' by zero in his *Līlāvātī*.

Bhāskara II's Concept of Zero

According to Bhāskara II, *śūnya* (zero) and *kha* (the sky) are synonymous. The logistics of zero[30] (*śūnya parikarma*) in *Līlāvatī* of Bhāskara II has mainly two aspects :

1. Taking *śūnya* as void, all the operations except division as defined by Brahmagupta have been corroborated.
2. In the interpretation of the case of division by zero, the word *kha* (the sky) and *śūnya* (used as synonymous) have definitely been used as a quantity existent in reality but negligibly small befitting the definition of '*tuccha*' (trifling or negligibly small) given in *Amarakoṣa* with the further improvement that this negligibly small quantity is of order zero.

According to John Taylor,[31] these lines mean a number divided a cipher has cipher from its divisor. When a number is multiplied by cipher, the product is cipher, but in case any operation remains to be done (*śeṣavidhau*), cipher is merely conceived to be the multiplier', i.e., if cipher becomes the multiplier and more operations remain to be done, cipher is to be considered as the multiplier without however performing actual operation'. Colebrooke[32] did not deviate from this meaning. According to Datta,[33] the term *kha* has been used here as an infinitesimal, a quantity which tends to zero. Datta's observation explains why Bhāskara has added:

śūnye guṇake jāte khaṃ hāraś cet punas tadā raśiḥ, i.e., a multiple of cipher is cipher, on division of the product by cipher, that quality is again (obtained).

DISCUSSION

Bhāskara II's concept of zero in connection with his interpretation of the operations of division and of multiplication by zero, is a quantity which actually tends to zero. The idea of tending to *śūnya*, a number existent in reality reminds us Nāgārjuna's complete *śūnyatā* or *nirvāṇa*, the ultimate mundane truth towards which all are moving seeking fulfilment. The concept of *śūnya* resembles Kaṇāda's concept of *ākāśa* in the sense of its singularity like sky (*ekatve cākāśasya*), its omnipresence among numbers (*yoge khaṃ kṣepasamaṃ, vargādau khaṃ, khenonitaś ca yutaḥ avikṛta eva jñeyaḥ*).

That the mathematical concept of *śūnya* in India is an outcome of Indian philosophical thoughts, will be clear if we think how the philosophy of *abhāva* had been almost a traditional guiding philosophy behind the pre-Bhāskarīya concept of *śūnya*. So, it is quite probable that Bhāskara II's concept of *śūnya* was built up against the background of Nāgārjuna's philosophy.

ACKNOWLEDGEMENT

The author expresses his indebtedness to Dr. M.R. Adhikari for his inspiration and also to INSA. authorities, New Delhi, for their kind co-operation in presenting this paper in the seminar on 'Concept of Śūnya', held during February 12-14, 1997 under the auspices of INSA and IGNCA, New Delhi.

REFERENCES

1. Sharma, Chandrakanta : *Vaiśeṣikadarśanam of Kaṇāda*, Sanskrit College, Calcutta, 1809, p. 156.
2. Raman, K. Venkata : *Nāgārjuna's Philosophy as presented in The Mahāprajñāpāramitā Śāstra*, Bhāratīya Vidyā Prakashan, Varanasi - 1, 1971, p. 31.
3. Raman, K. Venkata : *Nāgārjuna's Philosophy*, p. 30.
4. Bag, A.K. : *Mathematics in Ancient and Medieval India*, Chaukhambha Orientalia, Delhi, 1979, p. 26.
5. Datta, R.C. : *Ṛgveda Saṃhitā*, Haraf Prakasani, Calcutta, 1976, pages 132, 292.
6. Datta, B.B. : *History of Hindu Mathematics* (Part-I), Motilal Banarsidass, Lahore, 1935, p. 1.

7. Williams, Monier Monier : *A Sanskrit-English Dictionary*, Motilal Banarsidass, Delhi, 1986, p. 1085.
8. Nag, Surendra Pratap (Publisher) : *Atharvaveda Saṃhitā* (Śaunakīya), Nag Prakashak, Jawharnagar, Delhi, 1994 (Verse 14.2.19), p. 310.
9. Majumdar, R.C., Raychaudhuri, H.C. and Datta, K.K. : *An Advanced History of India*, MacMillan India Ltd., Madras, 1991, p. 40.
10. Majumdar, R.C., Raychaudhuri, H.C. and Datta K.K. : *An Advanced History of India*, P. 56, 86.
11. (Editor) Govind Krishnaji : The *Nāmaliṅgānuśāsana Amarakoṣa* with the commentary of Kṣīrasvāmin, Upāsana Prakāshan, Delhi, 1981, p. 174.
12. Samādhyayī, Sitanath (Editor) : *Piṅgala Chandaḥ Sūtram*, Calcutta, 1840, Śakāvda, p. 185.
13. Datta, B.B. : *History of Hindu Mathematics* (Part-I), Motilal Banarasidass, Lahore, 1935, p. 75.
14. Datta, B.B. : *History of Hindu Mathematics* (Part-I), p. 239.
15. Datta, B.B. : The Bakhshali Mathematics, *Bulletin, Calcutta Mathematical Society*, Vol. XXI, No. 1, 1929.
16. Dikshit, Sankar Balakrishna : *Bhāratīya Jyotish Śastra* (Part-I), Translated in English by R.V. Vaidya, India, Meteorological Department, New Delhi, 1981, p. 13.
17. Thibaut, G. and Dvivedi, S. : *Pañcasiddhāntikā of Varāha Mihira*, Benares, 1989, p. 54.
18. Bag, A.K. : *Mathematics in Ancient and Medieval India*, pp. 10-11.
19. Datta, B.B. : *History of Hindu Mathematics* (Part-I), p. 239.
20. Sarasvati, T.A. Amma : *Geometry in Ancient and Medieval India*, Motilal Banarasidass, Delhi, 1979, p. 9.
21. Sharma, Ram Swarup (Editor) : *Brāhmasphuṭa Siddhānta*, Indian Institute of Astronomical and Sanskrit Research, New Delhi, 1966, p. 243.
22. Sharma, Ram Swarup (Editor) : *Brāhmasphuṭa Siddhānta*, p. 244.
23. Sharma, Chandrakanta (Editor) : *Vaiśeṣika Darśanam* of Kaṇāda, p. 156.
24. Raman, K. Venkata : *Nāgārjuna's Philosophy*, 1971, p. 172.
25. Raman, K. Venkata : *Nāgārjuna's Philosophy*, p. 43.
26. Raman, K. Venkata : *Nāgārjuna's Philosophy*, p. 324.
27. Radhakrishan, S.: *Indian Philosophy*, Blackie and Son Pvt. Ltd., Bombay, 1977, p. 666.
28. Sharma, Chandrakānta (Commentator) : *Vaiśeṣika Darśanam* of Kaṇāda, P. 50-54.
29. Raman, K. Venkata : *Nāgārjuna's Philosophy*, p. 206.
30. Sarma, K.V. (Editor) : *Līlāvatī* of Bhāskarācārya with *Kriyākramakarī* of Śaṅkara and Nārāyaṇa, Vishveshvaranand Vedic Research Institute, Hoshiarpur, 1975, p. 91.

31. Taylor, John (Translator) : *Līlāvatī* or A Treatise on Arithmetic and Geometry by Bhāskara Ācārya, 1816, p. 29.
32. Banerji, Haran Chandra : *Colebrooke's Translation of the Līlāvatī*, The Book Co, Cal., 1927, pp. 24-25.
33. Datta, B.B. : *History of Hindu Mathematics*, p. 242.

APPENDIX

S-1 : *udva ūrmiḥ śamyā hantv āpo yoktrāṇi muñcata /*
māduṣkṛtau vyenasā'ghnyau śūnamāratāṃ //

Ṛgveda Saṃhitā 3.33. 13
(Ref. 5)

S-2 : *mo ṣu devā adaḥ svarava pādi divaspari /*
mā somyasya śambhuvaḥ śūne bhūma kadācana /
vittaṃ me asya rodasī //

Ṛgveda Saṃhitā 1.105.3
(Ref. 5)

S-3 : *uttiṣṭhetaḥ kimicchantīmāgāḥ ahaṃ tveḍe abhibhūḥ syād gṛhāt /*
śūnyaiṣī nirṛīte yājagandhottiṣṭhārāte pra pate mêha raṃsthāḥ //

Atharvaveda Saṃhitā, 14.2.19
(Ref. 8)

S-4 : *śūnyṃ tu vaśikaṃ tucchariktake*

(Ref. 11)
//dvir ardhe// 28

S-5 : *apanīte ityadhyāhāraḥ yadā jijñāseta saḍakṣare cchandasi kati vṛttāni bhavantīti? tadā tāṃ chandokṣarasaṃkhyāṃ bhūmau sthāpayitvā tato 'rdham apanayet/ tasminn apanīte dvau labhyate/ tatas tāṃ dvisaṃkhyāṃ bhūmau pṛthak prastārayet/ tataḥ śeṣāsyātra trayo'kṣarasaṃkhyā bhavanti/ teṣām ardhayitum aśakyatvāt kiṃ kartyavyam ity āha/*

(Ref. 12)
//rūpe śūnyam// 29

S-6 : *viṣamasaṃkhyāto rūpam apanayet, tasminn apanīte śūnyaṃ labhyate tat pūrvvalabdhāyā dvisaṃkhyāyā adhastāt sthāpayet tato dvisaṃkhyā avaśiṣyate, tato 'rddhe 'panīte punar dvisaṃkhyā labhyate, tāṃ śūnyādhastāt sthāpayet / tato rūpe śūnyaṃ labhyate tad dvisaṃkhyāyā adhastāt sthāpayet /*

(Ref. 12)

S-7 : *ṣaṭtriṃśatsaṃyuktā*
hyanalāṅkārkatrivargaguṇaśūnyaiḥ /
divasāḥ saptamagatyāṃ
cāro yastadvadaṣṭamyāṃ //

(Ref. 17)

S-8 : *dhanayor dhanam ṛṇam ṛṇayor dhanarṇayor antaraṃ samaikya khaṃ*

(Ref. 21)

S-9 : *dvividho hy abhāvaḥ yat kiñcin na kadācid bhavati gaganāravindādi, kiñcic ca kadācid bhavad api na rupāntareṇa na kenacid rūpeṇa kutracid bhavati, na sarvvadā khalv api sarvatra/ tatrādyo 'vastubhūto vastubhūtas tv anyaḥ*

(Ref. 21)

S-10 : *yan na bhavati tad abhūtaṃ yan na vidyate tan nāstīty anayos tulyārthaḥ/ prāg utpatteḥ parataś cātmaprahāṇān na bhavati kāryajātaṃ tadātmanetyabhūtam iti nāstīti cocyatye.*

(Ref. 21)

Philosophical and Mathematical Implications of Zero in Indian Culture

S.A. PARAMAHANS

It is a well accepted fact that mathematics has nourished almost all the branches of science and so the progress of science in a community may be measured by the amount of mathematization of science there. Hence, the progress of mathematics in a society can be taken as an efficient measure of the scientific development of a society. The concept of zero happens to be the most powerful tool in the realm of such an important discipline.

Although usage of *śūnya* may be found in Vedānta philosophy as is evidenced by some Upaniṣads[1] and the term *māyāvāda* is very well known for Advaita philosophy; the theory of nihilism as developed by Nāgasena (1 century B.C.) is expected to have its clues in the preachings of Buddha (born 562 B.C.); yet these concepts which simply tell the actual voidness of the apparent reality, hardly provide any genesis for the mathematical zero. Albeit, the great grammarian Pāṇini seems to have some clues for this concept. These have been discussed in detail by Mukherjee in his book.[2] But the zero of Pāṇini has nothing to do with the mathematical zero of place-value notation.

A caution for the *lopa* zero of Pāṇini: *lopa* is considered as one of the forms of zero given by Pāṇini. It is just subjective zero as relative zero of abstract algebra (which will be discussed later). Take for example, the case of liaison of *hare iha* and *viṣṇo iha*.[3] Here, after the *ayādeśa* or *avādeśa* of *e* or *o* the sūtra *lopaḥ śākalyasya*,[4] causes the *lopa* of *y* and *v* and so the occasion of a liaison takes place. But since this *lopa* happens to be a function initiated by *tripādī*[5] it becomes nullified due to '*pūrvatrāsiddhaṃ*'[6] and the liaison is avoided. In the case when the *lopa* does not occur, the forms of the compound words are *harayiha* and *visṇaviha* respectively. So here *lopa* of *y* and *v* is just subjective (it is just a view of Śākalya and his followers) and hence it may be termed subjective zero).

However, its origin in Indian soil as a mathematical concept is noteworthy as expressed by Halsted[7]: "The importance of the creation of the zero mark can never be exaggerated. This giving to airy nothing, not merely a local haibtation and a name of picture, a symbol but helpful power, is the characteristic of the Hindu race from whence it sprang".

It has four major phases or categories as enumerated by Gupta,[8] the first being that of emptiness or nothingness, the second being the concept of an indicator of the point of reckoning, measuring or graduation etc.

The third is the concept of zero as a symbol which denotes the absence of a denominational term in any positional numerical notation and the fourth is the concepts which treat zero as a number and which can normally be subjected to mathematical operations. In this connection he has mentioned the ancient mystical doctrine which looked at zero as a sort of infinitely small quantity which does not affect another quantity additively or subtractively, i.e. $N \pm 0 = N$.

It is to be pointed out here that abstract algebra uses the concept of zero in a different sense too. There the term 'small' is used in relative sense and has its synonym as 'superfluous'. Thus a non-trivial submodule N of a module M is small[9] or superfluous in M if $K = N = M$ implies $K = M$. So here a non-zero submodule N plays the role of zero. Amusingly, the set Z of all the integers (positive, negative and zero) is a small submodule of the set Q of rationals. Mathematically, $S + Z = Q$ implies $S = Q$ (where Z is a submodule of Q). Hence a module may be small in one module and may not be small in some other. Still more peculiar phenomenon is the following: All the non-trivial submodules or subgroups act as zero if a quotient is formed with respect to them. Mathematically, a non-trivial submodule N of a module M will be the zero of the quotient module M/N. Similarly, a non-trivial normal subgroup H of a group G is the zero of the quotient group G/H.

Another amusing concept is found in topology. The empty space does not admit even the numerical zero element in it.

It is very well known to historians of mathematics that systematic powers of 10 are found in *Taittirīya Saṃhitā*[10] up to 10^{19}. *Śatottaragaṇanā*, in which the next term is hundred times the preceding term, is found in

Lalitavistara.[11] It goes up to 10^{53} and *lakṣa* scale in which the next term is 10^5 times the preceding term and which gives terminology of numbers up to 10^{60}, is found in Vālmīki's *Rāmāyaṇa*.[12]

Although systematic powers of 10 and the *lakṣa* scale indicate the knowledge of zero, yet some scholars do not agree with it and assert that the presence of the names of powers of 10 in the *Taittirīya Saṃhitā* does not mean that the authors of that text used zero and similarly the terminology in terms of multiples of 10^5 proves nothing about the use of zero in India. But it is hardly believable to imagine the systematic multiples of 10 or 10^5 without the use of zero for place-value notation. Even ignoring the passages of *Taittirīya Saṃhitā* and Vālmīki's *Rāmāyaṇa*, the use of zero is found in the *Chandaḥsūtras* of Piṅgala[13] (approx. 200 B.C.) and *Bakhshālī Manuscript*[14] (Datta and Singh: *History of Hindu Mathematics*, Part I (1935), p. 75.

It should be mentioned here that the place value notation mentioned above is due to Indians as French mathematician Laplace remarks:[15] "It is India that gave us the ingenious method of expressing all numbers by means of ten symbols, each symbol receiving a value of position, as well as an absolute value; a profound and important idea which appears so simple to us now that we ignore its true merit, but its very simplicity, the great ease which it has lent to all computation, puts our arithmetic in the first rank of useful inventions; and we shall appreciate the grandeur of this achievement when we remember that it escaped the genius of Archimedes and Apollonious, two of the greatest men produced by antiquity".

Since the antiquity of zero goes at least up to the introduction of the place-value notation, it is worthwhile to investigate the time of this introduction. It appears that the ancient Indians introduced the decimal place-value after the early Vedic period because we find that the number 3339 has been mentioned by additive method in *Ṛgveda*,[16] as 300 + 3000 + 30 + 9 without using place-value notation. Even in *Mahābhārata*,[17] *Strīparva*, we find the same method to express as 16600200000 as 1000000000 + 20000 + 660000000 and 24165 as 24(1000) + 100 + 60 + 5.

But certainly the place-value notation was used before the early centuries of the Christian era since in the Kamalaśīla's commentary[18] on the *Tattvasaṃgraha* of Śāntarakṣita, we find a quotation from Vasumitra (A.D. 100) which tells that the place-value notation was employed in A.D. 100 for delineating philosophical principles. It is only possible if this notation was very well known in the public and so the period of its invention is expected to go centuries back before the Christian era. This fact is also supported by Dentzig:[19] "An inflexible numeration so crude as to make progress well-nigh impossible and a calculating device so limited in scope that even elementary calculations called for the services of an expert . . . . Man used these devices for thousands of years without making a single worthwhile improvement in the instrument, without contributing a single important idea to the system . . . . Even when compared with the slow growth of ideas during the Dark Ages, the history of reckoning presents a peculiar picture of desolute stagnation. When viewed in this light, the achievement of the unknown Hindus who sometimes in the first centuries of our era discovered the principle of position, assumes the proportion of a world event".

A firm evidence regarding the publicity of place-value notation in ancient India is *Vyāsabhāṣya*[20] on the *Yogasūtra* 3.13 as *yathā ekā rekhā śatasthāne śataṃ, daśasthāne daśa ekañ caikasthāne.*

It gives one more information: the author Vyāsa did not say *yathaiko aṅkaḥ* but *yathā ekā rekhā* which indicates clearly that the commentary of Vyāsa was written in a period when the numerals were written by strokes.

S.C. Kak explored the sign for zero and concluded thus:[21] "In view of the demonstration of the derivation of the Brāhmī numerals from the early Indus numerals[22] especially how the Indus 5 and 10 became the corresponding Brāhmī signs, the above derivation pushes the ancestry of the zero sign to the third millennium B.C." (Kak 1989, 1990).

To conclude, we see that the development of the concept of zero and its form in India were related to the form of the Brāhmi sign for 10. We have identified a late form of Brāhmi 10 as the immediate parent of the zero sign.

REFERENCES

1. *Upaniṣatsaṃgraha*: ed. J.L. Sastri, Motilal Benarsidass, Delhi (1980): *Muktikopaniṣad*, verse 2.55, p. 663 and *Tejobindūpaniṣad*, 1.41-1.42, p. 263.
2. R.N. Mukherjee, *Discovery of zero, and its impact on Indian Mathematics*, Das Gupta and Co. Pvt. Ltd., Calcutta-700073 (1991).
3. *Laghusiddhānta Kaumudī*, ed. Surendra Deo Snatak, Chaukhamba Orientalia P.O. Box 1032, Varanasi 221001 (1994), pp. 44-45.

4. *Aṣṭādhyāyī* of Pāṇini: Gokuldas Sanskrit series No. 29, Chaukhambha Orientalia, P.O. Box 1032, Varanasi - 2211001, Sutra 8.3.19.
5. *Ibid.*; *Sūtra* 8.2.1 — till the end of *Aṣṭādhyāyī.*
6. *Ibid.*; *Sūtra* 8.2.1.
7. G.B. Halsted: On the Foundation and Technique of Arithmetic, Chicago (1912), p. 20.
8. R.C. Gupta: Who invented zero? *Ganita Bharati* 17 (1995) pp. 45-61.
9. K.R. Goodearl, *Ring Theory*, Marcel Dekker Inc., New York and Basel (1976), p. 20, ex. 20.
10. *Taittirīya Saṃhitā*, ed. S.D. Satavalekar, Swadhyaya Mandal Paradi (1990), 7.2.20.
11. B.B. Datta and A.N. Singh, *History of Hindu Mathematics*, Part I, Motilal Banarsidass, Lahore (1935), pp. 10-11.
12. *Vālmikīya Rāmāyaṇa*, Gita Press, Gorakhpur (VS 2040), *Yuddha Kāṇḍa*, 28.33-38.
13. Ref. 11, p. 75.
14. Ref. 11, p. 77, Folio 56 verso of the manuscript uses zero in calculation.
15. Lancelot Hogben, *Mathematics for the millions*, George Allen and Unwin Ltd., London, 11th impression (Dec. 1938).
16. *Ṛgveda*, ed. Satavalekara, Swadhyaya Mandala, Paradi Balsad (1985): The hymn 3.9.9 reads as *trīṇi śatāni trī sahasrāṇy agnim trimśac ca devā nava cāsaparyan.*
17. *Mahābhārata*, Gita Press, Gorakhpur, III edn., vol. III (V.S. 2044), *Strīparva*, 26.9.
18. K.S. Shukla has informed in his Hindi translation (p. 72) of 'History of Hindu Mathematics (ref. 11 here)' that the commentary of Vasumitra on the *ślokas* 1787-1790 cites place-value notation as an example.
19. Ref. 15, p. 283.
20. *Yogasūtras* with the commentary of Vyasa along with *Tattvavaiśāradī* and *Yogavārttika*, ed. Narayana Misra, Bharatiya Vidya Prakasana, Varanasi, 1982, Sutra 3.13, p. 292.
21. S.C. Kak, The sign for zero; *The Mankind Quarterly*, 30(3) (1990), pp. 199-204.
22. S.C. Kak, Indus Writing, *The Mankind Quarterly*, 30 (1989), pp. 113-18; and S.C. Kak, Indus and Brahmi, Further Connections, *Cryptologica* 14 (1990), pp. 169-83.

Ramification of Śūnya in Other Fields

The Concept of Śūnya in Tantra and Āgama

H.N. CHAKRAVARTY

The concept of *śūnya* is difficult to explain and at the same time more difficult to convey the idea to others. The term has a number of synonyms, viz. *kha, nabhas, ākāśa, vyoma, randhra*, space, indicating hollow, barren, desolute, deserted, vacant and similar other phenomenal psychological and spiritual states. These terms only partially indicate what exactly the nature of *śūnya* is, because they fail to transmit the entirety of the concept. The doctrine speaks of *śūnya* as the Supreme Reality.

Āgama states *śūnya* as of partless form characterised by dissolution of all positive entities. It is sometimes described as the fire of the cremation ground. As a phoneme it is *ha* preceded by *sa*. The *Śardhakālottara* describes it thus: *śāntaṃ sarvaçataṃ śūnyaṃ mātrādvādaśake sthitam.*

The syllable *ha* preceded by the phoneme *sa* is of the nature of the void which abides in the end of the twelve *mātrās.*

The *Trayodaśakālottara* calls it *ākāśa*. It is void depending on none: *yac chūnyaṃ tan nirālambam ākāśaṃ parikīrtitam.*

It has used the term in two different ways, the one is the means and the other the end. To them, the supreme reality known as Śiva is *anuttara*, the beyond which is no other than *śūnya*. It is grasped and viewed in six different forms. They are the world, the embodied self, light (*bindu*), *śūnya*, the primodial sound and the dynamic *mantra*. In these above six aspects Śiva is to be seen, therefore Śiva is stated to be sixfold.[1]

Jayaratha while commenting on the above explains *śūnya,* the one of the sixfold forms, as of the characteristics of *śakti, vyāpinī* and *samanā*. These terms, viz. *śakti* and so on, I suppose, need a brief clarification. In the citation the term *kham* is called *śūnya* which includes *śakti, vyāpinī* and *samanā*. It is to be noted in this context that when the aspiring yogin is coursing along the path of development, he in the beginning has to traverse *bindu* located at the joint of the two eyebrows. It is ball-shaped like a *kadamba* flower — *kadamba-golakākāraḥ sphurattāraka-sannibhaḥ*, shining brilliantly like a sparkling star. Then onward when proceeding forward he has to pass through three voids known as *vyāpinī* and so on.

According to some tradition, the number of *śūnyas* are seven. Among them six are to be accepted in the beginning but ultimately they are to be rejected, while the seventh which is also of the nature of *śūnya* is acceptable to the seeker. These *śūnyas* are conceived by the yogin as subtle pulsation of the *mantra* constituents. The *mantra* of Tripura Sundarī is generally known as consisting of three sections. The third components of the above mantra are *bakāra, ikāra, ardhacandra, śakti, mahābindu* and the above — in the location of each of them six voids are to be conceived. The names, according to Amṛtānanda the commentator, are *repha, bindu, rodhini, nādānta, vyāpikā* and *unmanā*.[2] He says, "These three voids are to be known, o dear, located alternately. The location beyond these three voids is the great void. It is without any form (*nirākāram*). Other *śūnyas* are like the colour of feathers of the peacock.

From this discussion it is clear that voids are many. Some of them are attained first but finally they are to be rejected. It may be asked why the six voids are considered non-acceptable and why the seventh is finally acceptable. To this question, it is stated that the above six voids are relative by nature because they are moving, while the final one is absolute.

The Supreme Reality according to Śaiva doctrine is Śiva, who is transcendent, *anuttara*, that is, there is nothing beyond it. It is sometimes called *Śiva-vyoma*. Only from the following description we may be able to understand the concept.

The void called Śiva is beyond the grasp of *pramāṇa*, that is, it is beyond even the emergence of any instrument of knowledge, nor does it come under the scope of comparison. Because in comparison, we know that two entities are involved, the one is the thing to be compared and the other to which it is compared. Here distinction or difference indeed remains present. It is considered to be *ākāśa*. It is not the state of the physical void (*bhūtākāśa*) but even transcends the end of *cidākāśa*. It is *abhāva*, negation, the ultimate end. It is the cause of the pulsative stir (*saṃrambha*) of the blossoming of positivity. It should be known as the void of Śiva who is characterised as the base of all.[3]

The void is conceived not only as the basis of the entire superstructure of the universe but also the ultimate end of all. *Śūnya* is sometimes designated as all-consuming fire, in which all that is composed of parts are to be thrown to be burnt into nothingness so that one may be able to attain steadiness in the concept of the partlessness of the void (*niṣkalaṃ rūpam*).

In *Mahānaya Prakāśa* the void has been defined as the very body of the reality. It is stated that its specific form is *śūnyāśūnya*. While elaborating the idea the text states that *śūnyam* is composed purely of consciousness (*śuddhaṃ cinmātram*). It is not of the nature of *abhāva*, negation. Though it should be conceived as the void it is *aśūnyam* implying that it is not defiled by *māyā*.

The text very interestingly distinguishes the concept of various *ākāśas* and mentions three such *ākāśas*, viz. *bhūtākāśa, cidākāśa* and *parākāśa*. It says that the terminating limit of the universe, made of five gross elements is the physical void, *bhūtākāśa*. While this *bhūtākāśa* terminates in *cidākāśa*, the void of consciousness and that also goes back to the transcending void. This *bhūtākāśa* is not acceptable even though this physical void is the reflection of *cidākāśa*, because it does not have sentiency, while in the supreme void (*parāmākāśa*) both will and consciousness remain dissolved, *icchā citkramaḥ pralīnaḥ*. While in *cidākāśa* a radiation of will is possible, *cinnabhasi eva icchāprasāraḥ samasti, na jaḍe bhūtākāśe.* The commentary cites from an unknown source which states that from the void *prasara* pulsation/vibration occurs out of which the syllable *a* emerges and from that *a* the universe is born and from which the universe of the pentadic form comes into being.[4]

We have the similar idea in the Tantric text which states that the source of the universe is the syllable *a*. It is designated as *śūnya*, the void. In the syllabic form it is known as *akāra,* the first of the Sanskrit alphabets. The reason why it is called *śūnya* is that it is such a state in which the entire universe has gone to dissolve (*nirasta-nikhilaprapañcatvāt*). In order to understand the Tantric doctrine, it should be kept in mind that whatever becomes manifest it shines, at first with the aspect of consciousness known as *samvitkrama*. The next aspect is that of *dharmakrama*, the realm of light and lastly *varṇakrama*, the aspect of syllable. The syllable *a* which has been referred to here is the void. As such it is none other than the Supreme Śiva, so says *Saṃketa Paddhati*, an old Tantric text.

Akāra is the first of all phenomena who is the Supreme Śiva, the light.[5] In Śaiva doctrine it is known as *anuttara*.

The *Kiraṇāgama*, as we find quoted by Jayaratha in his commentary on *Tantrāloka*, states why Śiva is considered *śūnya*, void.

Why the principle Śiva is called the void? That void cannot be perceived. Perception is the means to arrive at perceptual knowledge and that which is beyond it is nothing.[6]

In reply the doubt is resolved in the following manner. Śiva is considered to be of the nature of the void because it is free from all attributes imposed over by *māyā*.

It is free from all *māyā dharma*. Being so it is known or conceived as *śūnya*. As Śiva transcends everything it is called the void. Moreover, it is *niṣkala*, partless when seen as the transcendent, the *viśvātīta*, but as immanent it is *sakala* which simply means with *kalā* parts or in other words it is possessed of *kalā* means *śakti* which is invariably associated with it.

Kālottara emphasises Śiva as without any content, which does not have any base to lend it any support. It is just like *ākāśa*.

The world as we see and feel does indeed have a form and name. But these forms are simply attributes, outer sheaths, so to say. When they are put aside, that which remains is without any real essence known to us as the void.

We fail to conceive the nature of void by following the course of negation so that we may arrive at the void described as nothingness, emptiness, devoid of any content. Whatever it may be it is only *parikalpita,* imagined. It cannot be the final resting ground, for if it is something thought of or conceived, the logic will naturally arrest our flight of imagination by saying that it is not the void but a non-void *aśūnyaṃ mātraṃ kevalam*, it is non-void and simply an idea.

The ancient thinkers in order to compromise between *śūnya* and *aśūnya* state that *śūnya* being of the nature of nothingness without any content in it, is of no use and for this reason it has no existence. Therefore it should be termed as *aśūnya*.

śūnyaṃ na kiñcit tac chūnyaṃ tv aśūnyam śūnyatā na hi/
yad akiñcit kathaṃ tadd hi na kiñcic chettum arhati//

Emptiness is nothing for it is void, that is empty of content. If it is said non-void, then indeed it cannot be called emptiness. What is negligible is unable to counter.

The above citation is from *Urmimahākaula.*

But it has been asserted by the Tantric teachers that both *ātman* and Śiva should be conceived as something without any content but the former should be conceived as the one who is free from all defilements in whom the divine qualities belonging to Śiva have not flourished yet. But *ātman* and Śiva are to be known as of the nature of *śūnya*, the void known as Śiva is highly attainable by *ātman*, the void.[7]

In this citation both *ātman* and Śiva are voids, the one is the means and the other, the end.

The great Śaiva exponent Ācārya Utpala in his *Pratyabhijñā Kārikā* states *śūnya* as absence of object of knowledge. While commenting on the above, Abhinava dilates the matter in a greater detail and finally defines what *śūnya* really means.

The ultimate reality according to Abhinava is *saṃvit,* consciousness. This *saṃvit* while going to cover its pure nature shines with the characteristic from intellect down to body and the jar. This indeed is one and manifests as such because of its play or freedom. Here no order of sequence or difference is real, but on account of freedom, difference and sequence may shine. This being the case, the portion of veil belonging to the principle of consciousness, itself shines without being mixed with any other state. It is such a state wherein the next state has not come into being, where all entities after being dissolved have just gone to rest. It is similar to *apavādya samādhi.* Here, in this very state the agenthood of doing (*kartṛtā*) primarily belonging to I-consciousness, abides indistinctly as a residual trace. As this is very unlike all phenomenal concepts known to the world, it does not have any form because of the absence of knowable object. With the result it becomes associated with negation without having any object to be known. The content of that concept shines as *na* nothingness as a residue. . . *saṃskāraśeṣībhūta jñeyarūpa*, that is called *śūnya.*[8]

Consciousness when becomes limited *saṃkucita* is called the void as it is of the nature of *bodha* consciousness, the question of its being called a knowable does not arise. But on account of being impregnated with doership it is supposed that this void is the impelling force that causes the vital energy to move —

tatraiva śūnyapramātari samavetā prāṇāpānasamānodānavyānātmakavāyucakre preraṇātmikā śaktiḥ sa ca vidyākālayoḥ prapañcabhūtau yau krameṇa buddhīndriya-karmendriyavargau tayor antarīvṛttiḥ.

In that very *śūnya* is the subject that is a power which is responsible for the working of the group of airs, called *prāṇa* and so on. This power is nothing but inner working (*antari vṛtti*) of the groups of senses of perception and those of reach which are mere expansion of *vidyā* and *kāla* respectively.

This body of ours though seen as composed of flesh and blood even then it is proper to view it as a funeral pyre, for everything gets dissolved there and the host of principles is seen burning in the flames are like all-consuming fire of death *Kālānala*.

This funeral ground is of the form of the void haunted by yoginīs and siddhas. This place of their sport is extremely dreadful because all forms whatever they may be, have become nullified. This *śmaśāna* is encompassed with the orb of its luminous rays by which great darkness has been dispelled. As it has got rid of all sorts of thought-constructs, it only shines in its refulgent light of blissfulness. Innumerable funeral pyres are seen burning here and there which evoke fear. These voids are blissful by nature *antaḥ svānubhāvānanda* and are experienced as one's own delight. It is free from defilement. It fills the universe — *viśvapūrṇam*. To the spiritual aspirants the meditation of these voids is highly important for it is the means for the attainment of the supreme and steadiness in the reality. We may cite a few voids here which serve this purpose. The one is *sūnyatā* and the other is *śmaśāna*.

yatra sarve layaṃ yānti dahyante tattvasañcayāḥ: /
tām citim paśya kāyasthāṃ kālānalasamaprabhām //
śūnyarūpe śmaśāne 'smin yoginīsiddhasevite /
krīḍāsthāne mahāraudre sarvāstamitavigrahe //
svaraśmi maṇḍalākīrṇe dhvaṃsitadhvānta santatau /
sarvair vikalpanirmukte ānandapadakevale //
asaṁkhyacitisampūrṇe śmaśāne citibhīṣaṇe./

— *Tantrāloka, XXIX, 182-185ab.*

The aspirant should meditate on the knowing subject, the basis of all thoughts as empty, similarly the objects of knowledge to be thought so and finally the instrument of knowledge should be meditated upon as voids. This practice when continued with steadiness, the one in the beginning remains in the realm of *vikalpa* fluctuating thought but in the long run this thought terminates in *nirvikalpa* which demolishes all the barriers of time and space.

For the development of consciousness voids are utilized. In the

Buddhist Tantric spiritual tradition, *śūnyatābhāvanā,* meditation on the concept of nothingness is essential for the spiritual seeker. *Sūnyatā* is defined as pure light shining in the form of *grāhyagrakadiśkāla-prapacvañcita prakāśmātratmakam sacarācaram viśvaṁ.*

The meditation of the void should be made in such a way as is free from all elements of thought of objectivity and the knowing subject but shining as a unique light of non-duality alone in which the entire mobile and immobile universe abides.

REFERENCES

1. *bhuvanaṃ vigraho jyotiḥ khaṃ śabdo mantra eva ca /*
bindunādādisaṃbhinnaḥ ṣadvidhaḥ śivc ucyate //
— *Tantrāloka* I. 63.

2. *śūnyatrayaṃ vijānīyāt ekaikāntarataḥ priye /*
śuñyatrayāt pare sthāne māhāśūnyaṃ vibhāvayet //
— *Yoginīhṛdaya,* 176

3. *pramāṇotthānarahitam upamābhedavarjitam /*
bhūtavyomapadātītaṃ cidvyomāntapadam param //
bhāvapratyayasaṃrambham abhāvaṃ paramā gatiḥ /
śivavyoma tu taj jñyeyaṃ sarvādhāratvalakṣaṇam //
— *Tantrāloka,* II, under 140.

4. *śūnyāt prasara utpannaḥ prasārād aḥ samutthitaḥ /*
akārād viśvam utpannaṃ viśvaṃ pañcātmakaṃ jagat //
— *Mahānayaprakāśa,* p. 29

5. *akāraḥ sarvavarṇāgryaḥ prakāśaḥ paramaḥ śivaḥ //*
— *Yoginīhṛdaya,* I, com under 11

6. *śivatattvaṃ kathaṃ śūnyaṃ tac chūnyaṃ nākṣagocaram /*
pratyakṣyaṃ cākṣavijñānaṃ tadatītaṃ na kiñcana //
— *Tantrāloka,* I, under 75

7. *ātmā śūnya iha jñeyaḥ śiva dharmair vinā kṛtaḥ /*
śivaḥ śūnyo 'dhigantavyo vimalo 'mūrtavigrahaḥ //
— *Tantrāloka,* II, under 48

8. *tatraiva cāhantārūpam kartṛtāyāḥ padaṃ parāmarśo 'sphuṭatvādarūpātmanā saṃskāreṇa śuddhena vedyapadavīm aprāptena yukto bhavati tadā saivāvasthā na ity eva parāmarśaleśa nipekṣitō. niṣedhya budhyādiviṣayā suspaṣṭa parāmarśa sambhedapi . . svīkṛta saranyākāra niṣedhyā ata eva samskāre śeṣikṛtā jñeyā śūnya ityucyate.*
— *Īśvarapratyabhijñāvimarśinī,* Vol. I, pp. 261-62.

Concept of Śūnya in Śākta Tantras

M.C. JOSHI

Of Sanskritic origin, the word *śūnya* is commonly used in the sense of cypher or zero. It is regarded generally to be a numerical term standing for total loss of quantity, but with its combination with other numerals indicates its quantitative position arithmetically in a specific context. In the Indian tradition, however, it has been employed in more than one sense. For instance, *śūnya* in the Sanskrit literature has been used to convey the idea of a vacant spot (*śūnya-sthāna*), a deserted village (*śūnya-grāma*), loneliness and helplessness as could be noticed in expressions like *sarvaśūnyadaridratā* equating abject poverty with total emptiness of life.

However, more significant is the employment of *śūnya* in Indian philosophical systems particularly by Mahāyānic Buddhists and Śākta teachers. A significant point which deserves to be noted in regard to the transformation of the property of *śūnya* in arithmetic and the Indian thought is the conceptual change from quantitative and qualitative proportions. The Buddhist concept of *śūnya* and *śūnyatā* has been variously interpreted by scholars and we now propose to see here as to how it has been employed in the Śākta tradition as an expression, concept or symbol.

From a general survey of Śākta-tantric scriptures and tradition it seems that the Śākta thinkers employed the term *śūnya* although not very commonly, to explain certain concepts pertaining to subtlety of a particular form of *Śakti* or the Supreme Mother and associated thought and ritual practices. Amongst the available Śāktic references connected with *śūnya*, perhaps the one which may broadly be placed to a period after the sixth century AD is the following extract from the *Devī Upaniṣad* or the *Devyatharvaśīrṣa*:

oṃ sarve vai devā devīm upasthuḥ kāsi tvaṃ mahādevīti /1//
sābravīt - ahaṃ brahmasvarūpiṇī / mattaḥ prakṛti puruṣātmakaṃ jagat / śūnyaṃ cāśūnyaṃ ca //2//
ahaṃ ānandānanandau / ahaṃ vijñānāvijñāne / ahaṃ brahmā brahmāṇī veditavye/ ahaṃ pañcabhūtāny apañcabhūtāni/ ahaṃ akhilaṃ jagat //3//

"When all the gods approached the Goddess and enquired as to who she is. She spoke (thus): I am the embodiment of Brahman; (and) from me (alone) the universe with the elements of Prakṛti and Puruṣa has emerged. I am (both) the void and the non-void and bliss and non-bliss. I am the Supreme Knowledge and the dimension beyond knowledge. I am knowable Brahman and unknowable (non-Brahman). I constitute the five elements and also their absence, I represent the entire universe".

Containing contradictory expressions, the passage cited above, in fact, reflects the idea of both the Nirguṇa and Saguṇa form of the Goddess, implying that the terms *śūnya* and *aśūnya* also need to be looked upon in the same context. M.M. Gopinath Kavijraj has attempted to explain the same phenomenon, while, analysing aspects of Śākta philosophy in the following words:

"It has already been observed that the appearance of the universe in creation (*sṛṣṭi*) follows upon the self limitation of the divine power and cosmic end in dissolution (*pralaya*) follows from the self-assertion of same power. After the period of the cosmic night is over the supreme will, in cooperation with the mature *adṛṣṭas* of *jīvas*, manifests, only partially as it were, the essence of the Self, whereon the self is revealed as limited. The appearance of limitation is thus the emergence of oneself known as *avidyā* or *jaḍaśakti*, called also differently by the name of void (*śūnya*), *prakṛti*, absolute negation, darkness (*tamas*) and *ākāśa*. This is the first stage in the order of creation and represents the first limitation imposed on Limitless".

Woodrolf, quoting famous commentary on the *Śāradātilaka* by Rāghava Bhaṭṭa, writes on the subject:

"She who is essential existed in a subtle (that is unmanifested) state, as consciousness, during the final dissolution (*yā anādirūpā caitanyadhyāsanā mahāpralaye sūkṣma sthitā*). It would be simpler to say that She is then what She is (*svarūpā*), namely consciousness but in the creation that consciousness veils itself. These terms "formless", "subtle", "dark" and "empty" all denote the same unmanifested state in which Śakti is in undistinguishable union with Śiva, the formless consciousness".

Professor Haeckel, however, tries to explain the term *void* (*śūnyatā*) more precisely stating that it is not nothing as some have supposed, but that which is like nothing known to us; the ultimate formless (*arūpa*).

The enlarged version of *Devyatharvaśīrṣa* (verse 24) further elucidates the Śākta idea of *śūnya* as could be observed from the following verse:

mantrāṇāṃ mātṛkā devī śabdānāṃ jñānarūpiṇī/
jñānānāṃ cinmayātītā śūnyāṃ śūnyasākṣiṇī//

"Vibrant in the hymns in the form of alphabet, meaning in the words, supra-consciousness in the knowledge and energy of emptiness in the voids. She is Durga, the Goddess beyond whom there is nothing".

In the texts like *Bṛahannīla Tantra,* we find a somewhat corroborative statement:

iyaṃ nārāyaṇī kālī tārā syāt śūnyavāhinī/
sundarī raktakālīyaṃ bhairavī nādinī tathā //

Noteworthy, however, seems to be *Mahākālasaṃhitā* which refers to the term *śūnya* in more than one context of Śāktism:

śūnye brahmāṇḍagole tu pañcāśacchūnyamadhyake //15//
pañcaśūnye sthitā tārā sarvānte kālikā sthitā //

The verse makes a reference to *śūnya* (void) as *Bramhāṇḍa*, then mentions fifty voids (*śūnya*) and five (*śūnyas*) serving as the abode of Goddess Tārā and thereafter the whole phenomenon is stated to terminate in the domain of Kālikā. Before we try to explain the above cited *śloka*, we should understand the subtle concept of goddess Kālī, whom the Śākta Tantras call the *Ādyā* or the primeval Goddess, the first amongst the ten *Mahāvidyās* in the light of the *Mahānirvāṇa Tantra* (34 iv) defining her:

kālasaṅgrasanāt kālī sarveṣām ādirūpiṇī/
kālatvād ādibhūtatvād ādyā kālīti gīyate //32//

The *śloka* clearly indicates that the goddess is called Kālī because she swallows the time (*kāla*) and creates timelessness which has been styled as *Mahāśūnya* and referred to as identical with *Brahman* in Śākta tradition:

brahmāṇḍarūpāyāḥ śaktiḥ parabrahmasvarūpiṇī/
cicchaktir iti vijñātā śūnyaṃ tasyās tu koṇakaṃ//

(*Śaktisaṅgamatantra*)

The expression '*sarvānte kālikā sthitā*' mentioned above also conveys similar idea about Kālī as the destroyer of *Kāla* i.e. the Time, thereby creating and occupying the *śūnya*. The goddess Tārā, the second *Mahāvidyā* or *Dvitīyā*, originally a Buddhist divinity of prime importance and adapted in Śākta-Tantrism as deity of *Kālikula,* is considered to be a modified from of Kālikā. In the tradition cited above, five *śūnyas*, mentioned as her abode, in fact symbolize destructed *pañca mahābhūtas* (elements) which in her iconography are depicted as five-skulls representing her crown (*śvetāsthipādikāyuktaṃ kapālapañcaśobhitāṃ*) on this account. In this connexion, we may like to point out that the five elements in the Śāktic creative process are expressed through the gross concept of goddess Śrī, as *pañca pretas*, sometimes called as five Brahmas, serving as the support of her throne. Goddess Tārā is, therefore, conceived of blue colour symbolizing infinity as against Śrī who is *sindūrāruṇa,* i.e., of vemilion complexion. Similarly, fifty *śūnyas*, which according to *Śākta* scriptures stand for fifty letters of Sanskrit alphabet from *a* to *kṣa*, forming as *muṇḍa-mālā* worn by both Kālī and Tārā iconographically, are rendered soundless in the process of dissolution and turn into *śūnyas*. These have also been equated with fifty-*tattvas* (*pañcaśad varṇasaṃketaṃ pañcāśat tat tvam uttamaṃ/*) as the forces of creation which became powerless in the withdrawal process. Therefore, the Śākta tradition (*Kāmadhenu Tantras*) about Tārā records, "transforming every thing into void She alone exists there" (*sarvaśūnyālayaṃ kṛtvā tatra caikākinī sthitā/*)

In the present context it may be relevant to refer to *Tāropaniṣad* which describes the cosmic form of the goddess Tārā in detail as can be seen below:

Oṃ tatsad brahma / tadrūpaṃ prakṛtiparāṅganābham/ tatparaṃ padaṃ mahat, satyam tadahaṃ, hriṅkāraṃ raktavarṇaṃ, mannābhiḥ strīṃkāraṃ piṅgalābhaṃ, hūṃkāraṃ viśadābhaṃ maddhṛdayarūpaṃ bhūmaṇḍalaṃ phaṭkāraṃ dhūmravarṇaṃ matkhaḍgaṃ, oṃkārajvaladrūpaṃ manmastakaṃ, vedā maddhastāḥ, candrārkānalā mannetrā, divānaktaṃ matpādau, sandhyā matkarṇau, saṃvatsaro madudaro, maddaṃṣṭrāpaṅktī matpārśvau, vārartavo madaṅgulyo, vidyā mannakhāḥ pāvako manmukhaṃ, mahī madrasanā, dyaur manmukhaṃ, gaganaṃ maddhṛdayaṃ, bhaktir mama carma, rasaṃ madrudhiraṃ, vānnaṃ vasāṃsī phalāni nirahaṅkārā asthīni, sudhā manmajjā, sthāvarāṇi madromāṇi, pātālādilokau matkucau, brahmānandam mannātyaṃ, jñānaṃ manmanaḥ, kṣamā buddhiḥ, śūnyaṃ madāsanaṃ nakṣatrāṇi madbhūṣaṇāni, etad vairāṭakavapuḥ majjalaṃ satvaṃ, bindusvarūpaṃ, mahākārasvarūpaṃ jyotirmayaṃ viddhi śiraḥ, ugratārāṃ mahogrāṃ nīlāṃ ghanām ekajaṭāṃ, mahāmāyāṃ prakṛti māṃ viditvā yo japati, madrupāṇi yo vetti manmantraṃ yo japati, madrūpakalpitāṃ yo japati, bhagaṃ bhajati nirvikalpasādhakaḥ sadā madrūpo bhavati. Sarvāṇi karmāṇi sadhyāni nirbhayo bhavati gurūn natvā stutvā vastrābhūṣāni dattvā imām upaniṣadvidyāṃ prāpya māṃ yo japati, sa jīvanmukto bhavati /

A perusal of the above shows that the description of the cosmic form of Tārā is an enlargement of her iconographic concept stating that void is her throne (*śūnyamadāsanam*). It reveals a significant conceptual aspect of *śūnya* serving as seat of the goddess Tārā iconographically which is regarded as a corpse (*śavāsīnā*) placed amid the burning pyres (*citāmadhyasaṃsthā*) *in śmaśāna*, the cremation ground, in Śāktism. Hence, *śūnya* also becomes a synonym and symbol of *śmaśāna*. Another dimension of the term under reference is indicated by the following extracts from *Śyāmārahasya*, a late medieval work

mahāśūnyena yat kṛtvā pūrṇas tiṣṭhati yogirāṭ
nirālambapade śūnye yatteja upapadyate /
tad dharmam abhyasen nityam dhyāyanti kulayoginām

The text thus directs that *yogins* of the Śākta (Tantrik) order that they should concentrate on the void conceived within their own body so that they may attain liberation. Similar practices have also been suggested by the texts on Buddhist *sādhanā* of Tārā in case of the yogis of Vajrayanic order:

tayā sa niruddha syāt / ataḥ prakṛtipariśuddhāḥ sarvvadhamām iti siddhaṃ / sarvvadharmmam prakṛtipariśuddhatā vibhāvya sarvvadharmma śūnyatāṃ vibhāvayet / tatreyaṃ śūnyatā/

Further, while a Vajrayanic devotee of Tārā contemplates on emptiness, i.e., *śūnya,* he is expected to recite the *śūnyatā* mantra according to systems of available Buddhist *sādhanās*. The non-Buddhist Tantric tradition also accepts this practice and the *sādhakas* worshipping the same goddess also follow it.

A further elucidation of contemplative system, with some modification mentioned in the Buddhist context of Tārā, is found in *Śrītattvacintāmaṇi,* a work of about the fifteenth/sixteenth century by Paramahaṃsa

Pūrṇānanda who belonged to Mymen— singh area of Bangladesh. The following verses in the text (Chapter XV) dealing with *bhūtaśuddhi* connected with the prevalent system (*Kuṇḍalinīyoga*) deserve attention.

ekībhāvaṃ tayoḥ kṛtvā jīvātmaparamātmanoḥ /
vicintya śūnyaṃ paramaṃ pārthivāṃśaṃ jale gataṃ //108//
jalañ ca tejasākṛṣṭaṃ tejorūpaṃ vicintāyet /
nirvāṇaṃ vāyunā vahni vāyurūpam vibhāvayet //109//
ākāśenāhṛtaṃ vāyuṃ ravaṁ vai manasi līyate /
mano 'haṁkāravaśagaṃ mahattve mahatāhṛtaṃ //110//
prakṛtau ca mahat tattvaṃ prakṛtiṃ cintayed atha/
prakāśarūpiṇīṃ nityāṃ brahmavidyāsvarūpiṇīṃ //111//
sahasradalamadhyasthaṃ śūnyarūpaṃ paraṃ padam //112//

The author tries to clarify the order of ablution in the process of meditation to have feel of the Eternal Light, the Goddess, personifying the supreme knowledge for a devotee. He indicates that a *sādhaka* should begin his contemplation by a feeling of combining *jīvātmā* and *paramātmā* within a *śūnya* to be followed in mental stages co-mingling of one *tattva* (element) with the other besides dissolution of *tanmātrās* to reach the Everlasting Light, the Śakti whose ultimate abode is *śūnya* (void) in the centre of *Sahasrāra*:

sahasradalamadhyasthaṃ śūnyarūpaṃ paraṃ padam//12//

It may be observed from the above noted verses that Paramahaṃsa Pūrṇānanda has conceived *śūnya* in two stages; that is, in the first stage with the indifferentiable combination of *jīvātmā* and *paramātmā* and at the last stage as a void representing *Nirguṇa-Paramātmikā*. Yet he introduces a third dimension of the void calling it *candrabimbasadṛśa,* i.e. looking like the orb of the moon and subtlest of the subtle (*sūkṣmātisūkṣme*), known as all pervasive area of darkness and *śūnya*, as the supreme illumination (*śūnya-prakāśamparam*), the knowledge of which keeps yama away from the *sādhaka*.

śūnyañ ca pratibimbacandrasadṛśaṁ sūkṣmātisūkṣamaṃ padaṃ
sarvaṃ vyapya tamomayam tu viditaṃ śūnye prakāśaṃ param/
dṛśyādṛśyavivādabhāgasakalaṃ jyotirmayaṃ sarvato
jñātvā tat padakam tu sādhakavarair dūrīkṛtaś cāntakaḥ //113//

The reason for such an explanation of *śūnya* with light may be on account of the fact that the author was a devotee of the Goddess Śrī, the Creative Energy according to Śākta concept of ten *Mahāvidyās* unlike Kālī and Tārā who are associated with the process of distruction (*saṃhāra*).

A point to be noted in the present context is that an early text on historical Śāktism like *Devī-mahātmya* does not mention the term *śūnya* at all in its narration although it does employ adjectives like Tāmasī (68 I) and citi (3 IV) (V perhaps to refer to the *avyakta* form of the Goddess. *śūnya* in the Śākta concept, therefore, may have been introduced subsequently, possibly not before the sixty century A.D. Most of the available Śākta-Tantras mainly use the term in context of Kālī and Tārā and in limited cases in relation to Śrī-Vidyā. It is tempting to state in the present context that perhaps under the Śaiva-Śākta influence an attempt has been made, as evidenced by a thirteenth century Sanskrit inscription from Vereval in Gujrat (AD 1264) to express the Islamic concept of God through the term *śūnya* in these words *namas te śūnya-rūpāya lakṣyālakṣya namostu te.*

I have tried to present our case on the basis of whatever source material we could collect on account of my own limited knowledge of the vast volume of Śākta Tantric scriptures. Therefore, my observations and inferences are to be treated as purely tentative.

Śūnya in the Context of Temple Art

DEVANGANA DESAI

The concepts of philosophy and religion find concrete and visible form in the architectural and sculptural imagery of the temple. The temple, as we know, is the abode of god and also a link between the divine and human planes. The temple is itself considered to be an image of the cosmos and also conceived of as Puruṣa, the personified image of the Universal Being.[1] It is a repository and synthesis of many symbols and metaphors.[2] In the temple we see a continuum of philosophy, religion and art.

How does the temple which is concerned with form (*rūpa, ākāra*) as well as dimensioned space, deal with *śūnya* (zero, void)? As we know, *śūnya* is a multilayered concept which has its place in metaphysics, mathematics, cosmology, science and other fields. In Yoga and with mystics, it is at the level of actual experience, not just at the level of an idea.

Among some of the cognate words for *śūnya* are *kha, ākāśa, nabha, bindu* and *chidra*. *Śūnya* is a term employed for the mathematical symbol zero, which is No-Figure and All-Figure. The word *pūrṇa* is also used for zero. As Ananda Coomaraswamy says, "We are immediately struck by the fact that the words *śūnya*, 'void'; and *pūrṇa*, 'plenum', should have a common reference, the implication being that all numbers are virtually or potentially present in that which is without number . . . . Again, employment of the term *ananta* with the same reference implies an identification of zero with infinity".[3]

Kha, which means undimensioned space, is also a term used for zero in mathematical usage of ancient India. Earlier, in the *Bṛhadāraṇyaka Upaniṣad* (V, I), *kha* is identified with *Brahman* and with *pūrṇa*. *Khaṃ*

Brahma, Khaṃ pūrṇam. The celebrated verse *pūrṇam adaḥ pūrṇam idam* in the *Bṛhadāraṇyaka* and *Īśāvāsya Upaniṣads* refers to the integral wholeness of the *Brahman.* All that emerges from it is also of the nature of all-inclusive Fullness. *Brahman* underlies the subtlest layers of existence. It is immeasurably small and infinitely vast. It has been likened to zero which symbolizes infinity.[4]

Indian mathematicians recognized that when a number is multiplied by zero it becomes zero and when divided by zero it becomes an infinity, *ananta, khahara.* It is significant that the concept of *khahara* is expressed in the form of a metaphysical analogy by Bhāskarācārya in his *Bījagaṇita* in the 12th century. In *khahara* there is no change whether much is added to it or substracted from it, just as no change takes place in the Infinite and Immovable (*ananta acyuta*) at the time of the emanation and resolution of the worlds, though hosts of beings are emanated or withdrawn from it.[5] Present day scholars treat the zero as infinitesimal, i.e. quantity decreasing indefinitely without actually becoming zero.[6]

The well-known simile of the *Brahman* likened to a spider (*tantunābha*) in the *Bṛhadāraṇyaka Upaniṣad* (I, 1.19), *Śvetāśvatara Upaniṣad* (VI, 10) and carried forth in the Purāṇas helps us to understand the notion of zero-point, which applies to the *bindu* in the *yantra* and the centre of *garbhagṛha* in the temple. The spider emits and reabsorbs threads (*tantus*) from its body. These threads extend in concentric circles and form wider circumferences. All the different points of the web are interconnected. "All of them can be traced back to the central starting point. This is the zero-point, which is the potential All-point". As Betty Heimann says, "*Brahman* is the productive zero-point infinitely extended and infinitely contracted".[7] Applying the simile of the spider to the *bindu* (point) of the yantra, Madhu Khanna says, "Like the spider in its web, the *bindu* (point) at the centre of the *yantra* is a centre of every creation, the radiating source of energy that generates all forms".[8]

Similarly, the nucleus of the *garbhagṛha* (sanctum) in the temple is the zero-point, the potential All-point. It is the metaphysical and physical centre of the temple. The *garbhagṛha* is a crucial area of the temple. It is a hollow chamber which is like a cave, *guhā*. It enshrines an emblem or the image of the divinity. Expansion proceeds from the central point of the *garbhagṛha* in horizontal and all directions of space.[9] It is a single principle from which everything issues and to which everything is connected. The finial, the highest point of the *śikhara* (superstructure), lies exactly above the centre of the *garbhagṛha*. This high point, the finial, is also conceived of as the visual equivalent of *bindu*.[10] The points — the centre of the *Brahmasthāna* on the ground level and the finial in the *ākāśa* (space, sky) — are connected by a vertical axis. As Stella Kramrisch says, it is an invisible axis which traverses the sanctum like a hollow reed (*śūnya madhya*). It is the Cosmic Axis which connects the heaven and earth.

It is on this physical and metaphysical point, *bindu*, of the *garbhagṛha* that the *liṅga*, described as *jagataḥ mūlastambha*[11] (primordial pillar of the universe) is installed. The *liṅga* is conceived of as *niṣkala*, undifferentiated, formless. It is the symbol of the Supreme Śiva, equivalent of the unmanifest Supreme Being. it has no *kalās* or parts. No measure can be applied to it. As T.A. Gopinatha Rao says, "In the *niṣkala* state the Supreme Being has no beginning, no limit or boundary and is pervading everywhere, extremely subtle and supreme and unknowable by any mode of proof".[12] The *Viśveśvara Saṃhitā* of the *Śiva Purāṇa* (V, 10) specifies that because Śiva is *Brahman*, he is *niṣkala, nirākāra nirguṇa*, i.e. undifferentiated, formless and without *guṇas* or attributes. Also he becomes Īśvara, who has *kalā* and form. As the Supreme Being he is worshipped as the *liṅga* and as Īśvara his *sakala* image is worshipped. The *Īśānaśivagurudeva-paddhati*, a Śaiva Siddhānta text, clarifies (Kriyāpāda, II, 36): *niṣkalaṃ kevalaṃ liṅgaṃ sakalaṃ pratimā smṛtā.* The *liṅga* is *niṣkala*, while *pratimā*, image, is *sakala*.

It is significant to note that it is only the *niṣkala* or emblem which is placed exactly in the centre of the *garbhagṛha* and not the *sakala* image. When the temple is dedicated to Viṣṇu, Sūrya, Jina or other divinities, the cult icon, *mūlanāyaka* or *dhruva-bera* is placed off the centre in the *garbhagṛha*. It is not in the exact centre as the *liṅga* is, but towards the back wall. The centre is kept vacant unless some symbol is put there. I am saying this on the basis of my observation of ground plans of medieval temples built in accordance with the *Vāstuśāstra* tradition. The temple architect Shri Hariprasad Sompura confirmed this practice in present times also and said that because the *liṅga* is conceived of as *svyambhū* (self-existent), it always occupies the centre of the sanctum, but not so the *sakala pratimās* of gods. The Indian *sūtradhāras* (architects) have thus distinguished between the positioning of the icon, *pratimā*, having form and the *liṅga*, sign of undifferentiated Supreme Being. The finial of the *śikhara* rises exactly above the *liṅga* but not exactly above the cult icon.

The installation of the symbol of the *niṣkala* exactly in the centre of the *garbhagṛa* with the Cosmic Axis traversing it from the finial above is the recognition of the pivotal significance of the undifferentiated, formless, unmanifest as the zero-point in the temple. Significantly, in a Khajuraho inscription (*Epigraphia Indica*, Vol. 1, verse 1, p. 140), Śiva is conceived of as *ajāyamāna bīja*, unborn seed of the universe in the form of a huge *vaṭa* (banyan) tree. It is with the *liṅga* in the centre that the centrifugal movement radiates in the four cardinal directions of the sanctum.

The unfoldment or evolution (*vistāra, prapañca*) from the central principle can be seen in the form of evolutes or manifested images, which are placed hierarchically in relation to the *liṅga* in some highly integrated and well-planned temples of medieval India. One such refined and mature temple, the Kandariyā Mahādeva of Khajuraho (c. A.D. 1030) represents the graded manifestations of Śiva in the iconic scheme.[13] As I am more familiar with the placement of images of this temple, I will briefly refer to it here. But the same can be said of the other mature temples, e.g. the Udayeśvara temple at Udayapur in Madhya Pradesh, the Mukteśvara and the Liṅgarāja at Bhubaneswar, or the Bṛhadīśvara temple at Tanjavur.

The visual expression of the doctrinal unfolding in the three stages from the unmanifest Supreme Śiva has been observed by scholars in several temples of India.[14] In the 6th century Elephanta Cave near Mumbai, Stella Kramrisch has pointed out the triple reality of Śiva in (1) the unmanifest *niṣkala liṅga* in the sanctum, (2) the manifest unmanifest or *sakala-niṣkala* Sadāśiva and (3) the manifest of *sakala* images or Śiva-Maheśa performing various *līlās* or sports, such as Dancer, Bridegroom etc.

In the Kandariyā Mahādeva temple of Khajuraho we observe the doctrinal unfolding from the *niṣkala liṅga*, to *sakala-niṣkala* Sadāśiva in the hall; and Maheśa's *līlā-mūrtis* in cardinal niches of the sanctum as well as in the hall. In addition, further evolutes Rudra, Viṣṇu, Brahmā and still lesser *aṃśas* or fractions in the rank of Śaiva hierarchy can be seen in the iconic scheme of the temple, arranged in graded sizes and different distances from the central *liṅga.*

The graded arrangement of images can be noticed, for intance, on the south wall of the sanctum. Śiva-Maheśa vanquishes the demon Andhaka in the main cardinal niche, which is topped by an image of Viṣṇu, smaller in size. Brahmā— still smaller— stands on the bottom row of the plinth.

In the context of this Seminar when we are discussing zero, fractions and micro-classificatory systems,[15] it is relevant to mention that the monotheistic Śaiva and Vaiṣṇava systems of medieval India have used fractions in their presentation of the emanatory process. The series of evolutes or divinities which emerged was bound to influence the hierarchy in the sizes and placement of images in temples.

I present here just a bare outline of the Śaiva emanatory scheme:[16]

Niṣkala Śiva—undifferentiated formless entity

|

Parā-Śakti emerges from the 1000th part

|

Ādi-Śakti emerges from the 1000th part of the previous one

|

Icchā-Śakti emerges from the 1000th part of the previous one

|

Jñāna-Śakti emerges from the 1000th part of the previous one

|

Kriyā-Śakti emerges from the 1000th part of the previous one

|

Sadāśiva—from the five *tattvas* which emerge from the 10th part of each of the five Śaktis.
(*sakala-niṣkala*)

|

Maheśa emerges from the 1000th part of the fifth aspect of Sadāśiva (*sakala*)

|

Rudra emerges from the 1000th part of Maheśa

|

Viṣṇu emerges from a crore part of Rudra

|

Brahmā emerges from a crore part of Viṣṇu

A series of evolutes emerge one from the other up the Pṛthvī *tattva*. In the involution process, all these *tattvas* are absorbed back into the primordial centre.

Thus in the Śaiva system each of the divinities is the fraction or *aṃśa* of the previous one and is diminutive *aṃśa* of the Ultimate Reality. Similarly, in the Pāñcarātra Vaiṣṇava system also there is a chain of emanations, "a process which, while bringing the product into existence, leaves the source of the product unchanged", as in *pūrṇam adah pūrṇam idam*.[17] The Supreme Being is the integral whole. Its essential nature is not affected by emanations.

The hierarchy and multiplicity of divinities gave rise to complex iconometric canons. Modular – fractional units for measurements were devised in religious and Śilpa texts.[18] *Tāla* and other units of measurements were employed to carve images into desired scales of proportions. Kapila Vatsyayan has pointed out the influence of algebraic system zero and place value on possibilities of permutations and combinations of the units of measures and proportions.[19]

Needless to say, the architectural expansion of the temple in the period required, apart from other technical know-how, the knowledge of mathematical calculations which must have been facilitated by the decimal place value system. It is significant that the 11th century Vāstu text *Samarāṅgaṇasūtradhāra* (Chapter 44) mentions knowledge of *Gaṇita* as one of the essential qualifications of the architect. This text gives elaborate proportionate measurements of various parts of the temple. Harmony attained by the choice of right dimensions is experienced in the presence of some of the magnificent temples of the 11th century A.D. at Udayapur, Modhera, Bhubaneswar and Tanjavur. Abstract principles of philosophy could be concretized with the help of mathematical calculations in the rhythmic movement of graded mini-spires, diminutive replicas of the principal spire, hierarchically spaced around the highest point the finial, of the central tower of the Kandariyā Mahādeva.

II

Briefly now we deal with *śūnya-pūrṇa* at the experience level.

The consecrated space of the temple, created by precise geometry and calculations, is designed for the manifestation of the Supreme Principle and also for the devotee to be able to "ascend" to the Primordial centre.

The Kandariyā Mahādeva temple, cave-like and mountain-like, is so designed that it takes the devotee upwards in his ascent towards the centre.[20] Climbing the successive levels of several *maṇḍapas* (halls) and moving in the interior space, passing by various *āvaraṇa* or surrounding divinities, he comes to the *antarāla*, the intermediary space, midway between the two units — the hall for devotees and the sanctum of the divinity. On the ground plan *antarāla* is "the incision which separates and at the same time connects *maṇḍapa* and *prāsāda*".[21] This unfilled space, the *antarāla*, is like a gap between the human level and the divine level. It joins them. On the exterior wall of the *antarāla*, some major temples of Khajuraho display the symbol of conjunction, the non-dual state of timelessness in which "there is neither day nor night, neither Sun nor Moon".[22] Standing at the *antarāla*, which is the *sandhi-kṣetra*, the devotee has *darśana* of the divinity.

This is not a place to talk on different approaches to the divine by devotees who may be at different stages of maturity. Some may attain the experience of *pūrṇa* through *bhakti* (devotion), while others may "respond to the proportions of the temple by an inner rhythmical movement", while the aspirant may sit in the *maṇḍapa* and perform certain rites of *antarayāga* described in Śaiva texts. The aspirant interiorizes the iconography through the yogic processes of *dhāraṇā* and *dhyāna* and abolishes the human condition.[23] Mentally going through different *tattvas* in an ascending mode of Brahmā, Viṣṇu, Rudra, Maheśa and then meditating on Sadāśiva on the point in the deep cavern (*bindu-gahvara*), he establishes his oneness with Sadāśiva. In the worship of *ādhāraśilā* performed before the worship of the *liṅga*, the aspirant invokes divinities in the involution order on a visualized eight-petalled lotus and worships Sadāśiva with *mantras* and *nyāsa* (ritual projection). Then he meditates on Śiva in the lotus of his heart, uniting him with Niṣkala Śiva.[24] When his own centre is stilled it coincides with the Still Centre, the Infinite, Immovable, the *pūrṇa*.

Without going into argument of diverse philosophical systems, it may be said that when one is free from the ego, *aham*, when one is "void" within, one can experience the state of *pūrṇa*.

The *Haṭhayogapradīpikā* (IV.56) expresses it very beautifully:

antaḥ śūnyo bahiḥ śūnyaḥ śūnyaḥ kumbha ivāmbare/
antaḥ pūrṇo bahiḥ pūrṇaḥ pūrṇaḥ kumbha ivārṇave//

"Within is void, without it's void, like an empty pot in space; completely full internally, completely full externally, just like a pot in the ocean".

I would like to end this paper with a vision of a Ṛṣi at the time between the Dissolution and Recreation of the universe. It is the profound theme of "Vaṭapatraśāyin", Viṣṇu-Nārāyaṇa on the banyan leaf floating on the Primeval Waters, a theme seen in later temple art and paintings and to which a temple is dedicated in the Bhakti-poet Āṇḍāl's village Srivilliputtur in Tamil Nadu. The subject is very powerfully expressed in the *Mahābhārata* (Vanaparvan, 188), *Matsya Purāṇa* (ch. 167) and the *Bhāgavata Purāṇa* (XI, 9). It takes us into the heart of *śūnya-pūrṇa.*

Viṣṇu sleeps like a spider who has drawn back the thread into itself. The god has consumed the web of the universe. The cosmos is within him. The manifest world has been absorbed in him. Only the sage Mārkaṇḍeya who has a boon of unending life roams endlessly in the universe inside the body of Viṣṇu. At one point the sage slips from the open mouth of sleeping Viṣṇu into the infinite sea, in the immense darkness. As H. Zimmer puts it, "He finds himself confounded by what amounted to Great Nothingness".[25]

At this time when *prakṛti* is *śūnya*, the sage has a vision of a radiant child sleeping on a banyan leaf. The divine child reveals himself to Mārkaṇḍeya as Nārāyaṇa, the Primeval Being, who is the source of the universe and who consumes the universe. The child opens his mouth and swallows Mārkaṇḍeya, who rests in him listening to the eternal rhythm of "*haṃsa*", the inhalation and exhalation, the breathing of the all-containing Universal Being.

It is significant that a medieval Pāñcarātra work quoted by F. Otto Schrader describes the divine child as *śūnya*.[26] "When the Day of the Lord has expired and the Great Dissolution is finished, nothing remains but the Waters of Infinity and, floating on them, on the leaf of a banyan tree (*vaṭapatra*), a babe whose name is "the Void" (*śūnya*). The babe is Viṣṇu, the sleepless one, sleeping the sleep of Yoga (*yoga-nidrā*)".

This is a beautiful description of the Supreme Being who has inhaled in all existence, the manifest world. He is self-contained, enclosed in himself like *śūnya* (zero). He is *pūrṇa.*

Myths have shades of meaning. The *Bhāgavata Purāṇa* (XII, 9) interprets the story as an instance of Viṣṇu's *māyā* (illusion). The Great Dissolution, *pralaya*, is not the actual cosmic dissolution, but it is an experience of a personal *pralaya*. The *Bhāgavata* describes the personal *pralaya* as the state of the severing of the bondage of the soul (Ātman), consisting of *ahaṃkāra*, which is an evolute of *māyā* and which comes in the way of the realization of the self. "Such a state of everlasting final emancipation is known as the eternal *ātyantika pralaya*" (XII, 4, 34).[27]

In his meditation, the sage Mārkaṇḍeya experiences the personal *pralaya*, when the bonds of *ahaṃkāra* are severed, when *prakṛti* is *śūnya*; and he has *darśana* of the Supreme Nārāyaṇa, who has the entire cosmos within himself, who is All-contained, *Pūrṇa*.

REFERENCES

1. Stella Kramrisch, "The Temple as Puruṣa", in *Studies in Indian Temple Architecture*, American Institute of Indian Studies, New Delhi, 1975. M. A. Dhaky, "Prāsāda as Cosmos", *Brahmavidyā*, The Adyar Library Bulletin, Vol. XXXV, 3-4, 1971, pp. 221-26. *Agni Purāṇa*, Chapter 61.
2. Kapila Vatsyayan, "Metaphors of Indian Art", The Sixty Nabadurga Banerji Lecture, March, 1996, *Journal of the Asiatic Society of Bombay*, Vol. 71, for 1996.
3. Ananda Coomaraswamy, "*Kha* and Other Words Denoting 'Zero', in connection with the Indian Metaphysics of Space", 1934, in *Coomaraswamy*, 2 : Selected Papers, ed., Roger Lipsey, Princeton University Press, 1977.
4. Betty Heimann, *Facets of Indian Thought*, London, 1964. Debarata Sensharma, in *Kalātattvakośa*, Vol. I, ed. Bettina Baumer, IGNCA, 1992, p. 434. Alain Danielou, *Hindu Polytheism*, London, 1964, pp. 6-7. "A Supreme Cause has to be beyond number, otherwise Number would be the First Cause".
5. Coomaraswamy, *op. cit.*, p. 226.
6. S. R. Sarma, in *Kalātattvakośa*, Vol. I, pp. 408 ff.
7. Betty Heimann, *op. cit.*, p. 99.
8. Madhu Khanna, *Yantra,* The Tantric Symbol of Cosmic Unity, London, 1979, p. 9.
9. Stella Kramrisch, *The Hindu Temple*, Calcutta, 1946, I, pp. 167, 179.
10. *Ibid.*, I, 353.
11. In Kālāñjara inscription of A.D. 1090, *Epigraphia Indica*, Vol. XXXI, pp. 163ff.
12. T.A. Gopinatha Rao, *Elements of Hindu Iconography*, 1916, Reprint 1971, Vol. II, ii, p. 375.
13. Devangana Desai, *The Religious Imagery of Khajuraho*, Franco-Indian Research, Mumbai, 1996, Ch. VI.
14. Stella Kramrisch, "The Great Cave Temple of Śiva in Elephanta: Levels of Meaning and their Form", in *Discourses on Śiva*, ed. Michael W. Meister, 1984. Doris Meth Srinivasan, "From Transcendency to Materiality : Para Śiva, Sadāśiva and Maheśa in Indian Art", *Artibus Asiae*, Vol. L, 1/2 1990.
15. Kapila Vatsyayan in *Journal of the Asiatic Society of Bombay*, Vol. 71, 1996, points out the simple equation 1-0/3 .333 gives rise to vast network of fractions. It is the bedrock of the principle of unity and diversity, one and multiple.
16. Based on T.A. Gopinatha Rao, *op. cit.*, Vol. II, ii, pp. 361 ff.

17. F. Otto Schrader, *Introduction to the Pāñcarātra and the Ahirbudhnya Saṃhitā*, pp. 39-40.
18. Bruno Dagens in *Kalātattvakośa*, Vol. II, pp. 377 ff.
19. Kapila Vatsyayan, in *Journal of the Asiatic Society of Bombay*, Vol. 71, 1996.
20. Devangana Desai, "Man and the Temple : Architectural and Sculptural Imagery of the Kandariyā Mahādeva Temple of Khajuraho", *Eastern Approaches*, ed. T.S. Maxwell, Oxford University Press, Delhi, 1992.
21. Stella Kramrisch, *The Hindu Temple*, p. 393.
22. Devangana Desai, "Placement and Significance of Erotic Sculptures at Khajuraho" in *Discourses on Śiva*, 1984.
23. Richard Davis, "Becoming Śiva and Acting as One in Śaiva Worship", in *Ritual and Speculation in Early Tantrism*, ed. Teun Goudriaan, 1992. Devangana Desai, "Man and the Temple . . . " in *Eastern Approaches*, 1992.
24. *Īśānaśivagurudevapaddhati*, II, xiii, 31-40, 50, 58-71. *Śiva Purāṇa*, *Vāyavīya Saṃhitā*, II, 24; *Kailāsa Saṃhitā*, Ch. 7.
25. Heinrich Zimmer, *Myths and Symbols in Indian Art and Civilization*, Bollingen Paperback, 1972, p. 46.
26. F. Otto Schrader, *op. cit.*, p. 99, mentions *Paramatattva nirṇayaprakāśa Saṃhitā*, 1, 1, 40.
27. D.F. Pocock, "Art and Theology in the *Bhāgavata Purāṇa*", *Contributions to Indian Sociology* (n.s.), 19.1, 1985.

Śūnya in the Indian Tāla System

PREM LATA SHARMA

Śūnya does not figure directly but indirectly in *niḥśabda kriyā* (lit. unsounded action) or silent units in the ancient *tāla* system and as a sign for *druta* or half *mātrā* in the medieval notation system. In contemporary Hindustani music, *khālī* (lit. vacant or empty) is an important component of almost all *tālas*. In notation it is represented by a zero. This current parlance has a long history.

According to Bharata's *Nāṭyaśāstra* (XXXI, 30, 31), *kriyā* or action in *tāla* is twofold, viz. *saśabda* (sounded) and *niḥśabda* (unsounded); each one of these two is again fourfold, viz. *āvāpa, niṣkrāma, vikṣepa* and *praveśa* are the four varieties of unsounded beats and *śamyā, tāla, dhruva* and *sannipāta* are the four types of sounded beats. These actions are further described as follows: —

āvāpa	(lit. sowing seeds) = holding the palm upwards and contracting the fingers inwards. This involves a movement of the hand towards the body.
niṣkrāma	(lit. exit) = holding the palm downwards and opening out the fingers. This involves a movement away from the body.
vikṣepa	(lit. throwing) = throwing the palm towards the right.
praveśa	(lit. entry) = holding the palm downwards and contracting the fingers. This involves a movement of the hand towards the body.
śamyā	= striking of the right palm on the left one.
tāla	= striking of the left palm on the right one.
dhruva	(lit. constant) = snap with the middle finger and thumb.
sannipāta	(lit. well-struck) = striking together both the palms held sideways, in front of the centre of the body.

In the ancient *tāla* system, there are five *tālas* and each one operates in three spans known as *mārga* (lit. path). The first span is composed of only sounded beats which are of unequal duration, with one exception noted below. The next two spans are double in duration successively. The three units of duration are :—

1. *laghu* = one *mātrā*. The notational sign is 1
2. *guru* = two *mātrās*. The notational sign is S
3. *pluta* = three *mātrās*. The notational sign is Ś

Mātrā is the smallest unit both in *chanda* (metre) and *tāla* that is equal to one short syllable in metre and to five short syllables or five winks of the eye in *tāla*. Since *tāla* operates in music, the duration of its basic unit is five times that of the duration in language.

The composition of the five ancient *tālas* in the basic span is as follows: —

1. *caccatpuṭaḥ*—SSIŚ = eight *mātrās* or four *kalās*.
2. *cācapuṭaḥ*—SIIS = six *mātrās* or three *kalās*.
3. *ṣaṭpitāputrakaḥ or* ŚSISIŚ = twelve *mātrās* or six *kalās*. *pañcapāṇi* or uttara
4. *sampakveṣṭākaḥ* ŚŚŚŚ = twelve *mātrās* or six *kalās*.
5. *udghaṭṭaḥ* SSS = six *mātrās* or three *kalās*.

Kalā is a standard unit of two *mātrās*.

Each of the first four of the above *tālas* makes a pattern of its own with its uneven components which are all marked with sounded beats. Only the last *tāla* has three equal components and hence its first unit is assigned an unsounded action in order to make a pattern. The span of these *tālas* is doubled by doubling the duration of each component unit but the number of sounded beats remains unchanged. Unsounded actions are prefixed to each unit, according to its duration as follows:—

First Double Span

1. *Laghu* is doubled as *guru* and hence no unsounded action is to be prefixed.
2. *Guru* is doubled as two *gurus*, the first of which is assigned the unsounded beat, *niṣkrāma* or *vikṣepa*.
3. *Pluta* is doubled as six *mātrās*, i.e. three *gurus*, the first two of which are assigned the unsounded beats, *niṣkrāma* and *vikṣepa*.

Second Double Span

1. *Laghu* is quadrupled as twc *gurus*, the first guru being assigned the unsounded beat, *niṣkrāma* or *vikṣepa*.
2. *Guru* is quadrupled as four *gurus*, the first three being assigned the unsounded beats, *niṣkrāma, vikṣepa* and *praveśa*.
3. *Pluta* is quadrupled as six *gurus*, the first five *gurus* being assigned the unsounded beats, *āvāpa*, *niṣkrāma*, *vikṣepa*, *praveśa* and again *āvāpa*.

The notable point here is that the extension of sounded beat is prefixed to it in the ancient system, whereas in today's practice, both in Hindustani and Carnatic music, it is suffixed, i.e. it follows the sounded beat that does not precede it. In other words, the subtle or abstract precedes the concrete in the ancient system but in today's practice this order is reversed and the concrete precedes the subtle or abstract. The ancient system appears to follow the Vedic concept according to which *sat* (existent) is born out of *asat* (non-existent) (*asataḥ sajjāyata, Chhāndogya Upaniṣad* VI.2.1).

In arithmetic, the zero succeeds a number and adds to its value. If it precedes the number, it adds nothing. Since *tāla* is a repetitive pattern and repetition operates in a cycle, the preceding unsounded extension of the first sounded beat could as well be taken as the succeeding extension of the last sounded beat. Coming back to the Vedic concept of *sat* (the existent) having come out of *asat* (non-existent), there are passages stating that *sat* was the first manifestation (cf. *Chhāndogya-Upaniṣad* VI.2.1)

sad eva saumya idam agra āsīt

The same *Upaniṣad* also says that there was *asat* in the beginning

asad eva idam agra āsīt (III.19.1)

In Hindustani music today, *khālī* comes in the middle of the larger unit (s), e.g.

1. *Teentāl* — The second unit is the double of the first and third units. Hence, *Khālī* falls in the middle of the second unit

1 2 3 4 / 5 6 7 8 9 10 11 12 / 13 14 15 16 /
C

2. *Cārtāl* — In this *tāla*, the first and second units are larger than

the third and fourth ones and hence, *khālī* falls in the middle of the first two units

1 2 3 4 / 5 6 7 8 / 9 10 / 11 12 /
0

In Carnatic music, the extension of a larger unit in a *tāla* is indicated not with a wave of the hand but in a different manner, e.g. in *Āditāl*, the first unit is larger and hence it is shown with the fingertips meeting the thumb. Therefore, sometimes it is said that there is no *khālī* in Carnatic music. But this is a misnomer. The phenomenon of one or more of the units of a *tāla* being larger than the others is very much there, although an equivalent term for *khālī* does not exist today.

Druta or half-*mātrā* unit was introduced in the *deśī tāla* system around the 5th century A.D. It was represented in notation with a zero. *Saṅgīta-ratnākara* V. 257 gives four names for *druta-ardhamātrā, vyoma* (space), *vyañjana* (consonant in language) and *bindu. Bindu* here stands for zero. *Śūnya* as a symbol of space and infinity (*kham*) is well-known in Indian thought.

"Employment of the term *ananta* with the same reference implies an identification of zero with infinity; the beginning of all series being thus the same as their end". (Ananda K. Coomaraswamy, "Kha and other words Denoting 'Zero', in connection with the Indian Metaphysics of Space", *Selected Papers—Metaphysics*, ed. Roger Lipsey, p. 220.)

In the domain of sound, silence plays a very important role which can be equated with the role of zero among numbers. The ancient *tāla* system highlights this importance of silence by juxtaposing *niḥśabda kriyā* with *saśabda kriyā* (sounded action). Since incessant action cannot bring about the perception of time, the interception of silence among sounded beats is essential for constructing *tāla*, the configuration of time in music. The hangover of this concept can be seen even in the contemporary *tāla* system. Whether silence precedes sound or succeeds it is immaterial in the domain of time which is conceived and perceived as cyclic in the *tāla* system.

Number Mysticism in Other Regions and the Impact of Śūnya

P. MANANSALA

In ancient times, the mathematician, in addition to often being astronomer, navigator etc., also played the part of priest and philosopher. Even today, the astronomer still functions as cosmologist. The ancient savants often related numbers as symbols, types, forms or even the embodiment of the various cosmological phenomena. Considering the fact that philosophers often see the greater cosmos harmonically mirrored in microcosm, the mysterious world of numbers with its series, progressions, harmonies etc., would naturally attract their curiosity and study.

The beginning, or the creation of the cosmos is one of the mysteries that has most intrigued the human mind. In relating the system of numbers to this beginning, the most common starting point before the advent of *śūnya* was naturally the number one. What could better represent the first cause. The number two, it follows, would naturally represent the beginning of duality.

There is a weakness in this system regarding cosmological views that view the beginning as essentially empty, or as an absence of objectivity. Eastern systems such as those of the Upaniṣads, Buddhism and Taoism essentially had such an outlook. The ideas of non-duality, monism, *śūnyatā* and similar concepts were well-suited, in regard to number mysticism, to the numeral *śūnya*, or zero.

Whether such philosophies, which predated the first recorded instances of the use of *śūnya*, had any influence on the development of the latter is beyond the scope of this work. Certainly it is possible, but whatever the case, the creation of a symbol for *śūnya* could not have found a better suited environment in which to arise. *Śūnya* was the missing piece in Eastern number mysticism.

SOME NUMBER MYSTICISM SYSTEMS OUTSIDE INDIA

Types of number mysticism appear in the West in the Egyptian and Chaldean cultures. Among the Egyptians, the eye of the falcon god Horus, known as *wdt* was seen as forming a decreasing geometric progression of fractions from $\frac{1}{2}$ to $\frac{1}{64}$ ($\frac{1}{2}$, $\frac{1}{4}$, $\frac{1}{8}$, $\frac{1}{16}$, $\frac{1}{32}$, $\frac{1}{64}$). This, of course, is the most simple and primal of progressions by which one divides into two, two into four, and so on.

Probably the Greeks picked up their initial number mysticism from Egypt, Chaldea and India by way of such early philosophers as Pythagoras, who is said to have travelled widely in search of knowledge. Pythagoras associated numbers with geometric figures and both numbers and geometric figures with the phenomena in the universe.

To the East, the ancient Chinese used numbers in the form of trigrams to express cosmic phenomena at a very early date. At the time of King Wen and the Duke of Chou during the Chou Dynasty (*c.* 1150 AE), the Chinese developed their intricate divination and cosmological system known as the I-Ching. According to Confucius, earlier forms of this book existed during the Hsia Dynasty (2205-1766 AE) and during the Shang Dynasty (1766-1150 AE). This sytem was based on 64 hexagrams occurring from all permutations of eight basic trigrams. The Chinese had a way of expressing zero by means of empty spaces on their counting boards.

Many other cultures also developed such types of number mysticism. Among the Maya, the number probably played more of a part in religion than anywhere else in the world. Mayan thought regarding numbers was connected with their unusual practice of viewing time as divine entity. Units of time were considered holy and gods were represented by numbers. Often the glyph representing a deity was the same as that of the deity's symbolic number. The coincidence of different units of cycles of time are considered very holy. Interestingly, the Mayan also had a method of expressing zero.

The ancient Babylonians also applied mysticism to numbers as did their preedcessors, the Sumerians. Assigning letters to numbers and deriving special mystical meanings from words in holy texts was a practice used by the Chaldeans and probably the Jews borrowed this practice from them. The Sumerians viewed the cosmos before creation as a great empty primordial sea (*abzu*). Around the fourth century AE, the Babylonians developed a method of representing zero in the positional number system they borrowed from the Sumerians. They inserted a space in the place of zero to indicate: 0 (place value).

EARLY CONCEPTS OF EMPTINESS IN COSMOLOGICAL THOUGHT

In the I-Ching, which is generally dated to about the end of the second millennium BE, each of the series of 64 hexagrams is made up of six places. In each of these six places is a symbol, either a line or a broken line. Graphically these lines can represent either one (the unbroken line) or two (the broken line). In the traditional process of divination with yarrow stalks, the unbroken line is generated by an odd number, while the broken line by an even number. Thus, we have positive-negative polarity between the two lines that can represent the whole gamut of male-female, odd-even, hot-cold and similar dualities.

In addition to polar opposites, the I-Ching system also generates an alternative to either the unbroken or broken line. When the number nine is derived from the calculations involving the yarrow stalks, the positive unbroken line becomes what is called the *old yang*. The term *yang* is, of course, the Chinese name for the positive principle. However, here the *old yang* becomes a changing line that acts not only as an unbroken line but also as a broken one. When the yarrow stalks generate the number eight, the broken line likewise becomes the *old yin* and acts both as a negative and positive line. In this sense the adjective *old* means primordial, and the *old yin* and *old yang* are examples of non-duality, or of the union of the two opposites. In the beginning, *yin* and *yang* are not divided and there is an absence of polarity. In mathematics, the merging of opposites results in zero: $1 + (-1) = 0$. Likewise, a quantity negated by an equal quantity equals *śūnya*: $1 - 1 = 0$. In more philosophic terms, me minus myself equals nothing; that is I return to my original state, which is relative non-existence.

The union of *yin* and *yang* represents both the beginning and the end. In the beginning there is no division, but then the changes develop and an interaction of *yin* and *yang* brings about various transformations which eventually result in a reunion of the two into non-polarity. The I-Ching, or Book of Changes, gives the order of these transformations as a model of the working of heaven and earth. Each hexagram is made up of two of a total of eight trigrams. The three lines in each trigram may represent the "three powers" mentioned in the *Ta Chuan*, a treatise on the I-Ching. Like the lines themselves, the three powers would be the positive (+ 1), the negative (–1), and the neutral (0). Here, again, *śūnya* can represent either a merging of opposites + 1 + (–1) = 0 or a simple absence of opposites (1 – 1 = 0), which in philosophic thought are usually considered to be the same.

From the interaction of the positive and negative powers, the eight trigrams, representing the eight Chinese seasons are formed. Two trigrams are paired together to form the 64 hexagrams and so from the basic polarity of the lines, a more complex polarity arises in the hexagram. In what is known as the Fu Hsi sequence, the hexagrams are ordered in a binary sequence that was later adopted by Leibniz. The sequence which, by the way, displays a startling resemblance to the codon sequence in the human DNA molecule. However, the trigrams and hexagrams in these sequences are formed only by the interaction of positive and negative. In the actual division, the netural, or changing, line also comes into play. The changing line is read as both as a *yin* and *yang* line. The order in which it gives such *yin* and *yang* characteristics depends on whether the line was generated by an odd/positive or even/negative number.

As the I-Ching is designed to examine the progression of changes in the cosmos, the fact that the neutral line gives both the attributes of *yin* and *yang* means, in effect, that absence of polarity always refers equally to the union of both polarities. In other words, absence, emptiness and similar concepts are not without effect or identity. Emptiness or nothingness is then rather fullness or the primordial state of all things. It is inclusive of both *yin* and *yang*, while each of those elements separately naturally excluded the other at the basic level.

Because the Chinese apparently had this idea of absence of polarity as an entity, it is not surprising that in their ancient system of counting rods they actually left an empty space as a zero marker. We can find that in many of the cultures that represented *śūnya*, there is a belief in a primordial emptiness from which all things arose. Nothingness has an identity and thus it becomes natural that it be represented symbolically. In cultures in which emptiness was just that, it might not be so natural to think of representing it at all. In later China, the ideas of the early sages eventually developed into Daoism, a system in which life and death, good and evil and all other opposites originate and merge back into the Dao. In the Dao, *yin* and *yang* are not differentiated.

In India, the first developed ideas of monism and non-duality appear in the Upaniṣads, the earliest of which were written a few centuries before the time of the Buddha. Brahma is stated to be the ultimate reality without a second. Individual reality is the same as the ultimate reality, a concept similar to that in Daoism. Brahma, or other forms of the Supreme Reality, are neuter in their absolute form, while they may have gender/polarity in their real form. Here the beginning goes back to a formless, attributeless God without division or polarity. From the one springs many (*ekānekarūpa*) in the real manifestation, but eventually all unite back with the one absolute reality. Such ideas were eventually developed by Śaṅkara into the doctrine of unqualified monism, or *advaitavāda* (doctrine of the non-second).

In Hinduism, these monist ideas competed with the concepts of *prakṛti-puruṣa* and *aja ekapada-ahir budhnya*, which saw the two poles as always existing and as exclusive to each other.

The rise of Buddhism saw the idea of the non-dual developed to a stage where it is seen as complete non-existence and emptiness. The Buddha himself is said to have stated, "There is, O monks, a non-born, a non-becoming, a non-caused. If there were not, there would be no refuge for that which is born, becomes, is created, is caused". The Ultimate Reality is known as *Nirvāṇa* and it is to this we all return: "Of him who has disappeared there is no measure. That by which they say he is, exists for him no longer". The idea of the "non-born" is similar to that of *aja ekapada*, the Unborn, who exists prior to the creation as a form of positive cosmic energy, but here there is no dualism involved.

From this seed planted by the Buddha, there developed in Mahāyāna Buddhism the concept of *śūnyatā*. Derived from the same root as *śūnya*, this term meant simply emptiness or voidness. This quote from the *Prajñā Pāramitā Hṛdaya Sūtra* gives some idea of this concept in Buddhism:

> *O, Sāriputra, remember, Dharma is fundamentally* śūnyatā, *no birth, no death. Nothing is defiled, nothing is pure. Nothing can increase, nothing can decrease. Hence, in* śūnyatā, *no form, no feeling, no thought, no volition, no consciousness. No eyes, no ears, no nose, no tongue, no body, no mind. No seeing, no hearing, no smelling, no testing, no touching, no thinking. No world of sight, no world of consciousness. No ignorance and no end to*

ignorance. No old age and death and no end to old age and death. No suffering, no craving, no extinction, no path, no wisdom, no attainment.

Śūnyatā becomes identified with the Buddhist concept of non-self and also with self. The real world and the real self are a result of *māyā* (illusion) and the Ultimate Reality is characterized as emptiness and the non-self. In the beginning there exists only *nirvāṇa*, or *śūnyatā*, which lacks form or reality in the human perception. From this arise the other beings. In the Mahāyāna school, this early nothingness was called the Dharmakāya Buddha. Thus, it was given an identity. Eventually, the Dharmakāya Buddha takes on real attributes and a personality developing into a type of Supreme God. One of the names given to this deity was naturally, *Śūnyatā*.

In Buddhism the real *śūnyatā* is found at the point of the "coincidence of opposites". It is only here that the opposites manifest their true identity, a process of "identity by negation". *śūnyatā* is viewed in its primary state as "self-emptying", it is not without "being". The existence of original *śūnyatā* is premised on primary unity and non-existence of opposites, which is the true being. Emptiness is seen as the true form, as this indeed is the original form, and the form to which all things are destined to return. The emptiness of *śunyatā* is not emptiness in absolute terms, which would mean the existence of nothing, real or absolute. Instead, it seems to represent the non-existence of that which we consider real, of that which divides us from our true being. Not surprisingly, in systems which have such concepts, a notion of pantheism, where all things are part of a great single divine whole, is very common. Thus, rocks, rivers, mountains, fish, insects, humans, stars and all other things are part of one great organism. Our seeming differences in the real world are illusory due to natural immutable rules which govern the cycles of existence.

Śūnyatā is that which existed before all else and which would necessarily preclude beings and objects familiar to us. That such a primordial state would be conceived of as emptiness is quite logical. The idea of returning to this state, so often equated with bliss, is the object of the esoteric thinker, who is prone to seeing all things working in cycles. From *śūnyatā* we arose, to *śūnyatā* we return. So, this concept cannot be classed correctly as nihilism. As the Buddha himself stated, the death of persons, and their passing on to *nirvāṇa* does not mean that they discontinue to exist. When the illusory self is negated, it goes back to its true identity where being goes beyond what our senses are able to comprehend.

The numeral zero also represents that from which all numbers arise, and by negation, that to which all numbers return. So *śūnya* can be equated with the real self, the primordial beginning in which no division exists, and the final end in which all things reunite. The positive and negative forces, represented by the sets of positive and negative numbers both have their starting points at *śūnya*. When united, these sets negate each other and return to *śūnya*/zero. When we look at the set of numbers as finite and circular, as in non-Euclidean geometry, then the finite real universe starts and ends on the same plane, which can be equated with zero.

So in India, the concept of emptiness also attains identity, even attaining to the status of the supreme deity. While associated with formlessness in the esoteric discourses, *śūnyatā* was depicted artistically in anthropomorphic form; and the neuter Brahma was also depicted in his real form. So it should not be so surprising that the zero of mathematics eventually came to be represented graphically. In Buddhism, negation of the self leads to *nirvāna*, which itself means to "extinguish a flame". The self minus the self comes to equal the Ultimate Reality (1 – 1 = 0), known as *śūnyatā*.

In Tantric Hinduism, while the square was used to symbolize creation, the *bindu*, or dot, represented the final dissolution when all things returned to their original unity. In some ancient texts, the *bindu* seems also to have been used to represent zero along with the *śūnya* symbol, which eventually took that role entirely. Like the Buddhists who sought *Nirvāṇa*, the Tantrics saw the chief goal of their efforts in achieving the supreme unity of the *bindu* state. Thus, something analogous with the dissolution of the universe was viewed as the goal of the devotee. Duality is the ultimate illusion that prevents one from unifying with the Void (*śūnya*), the original Śiva-Śakti union (the cosmic male and female principles). The primordial cosmic principle is known as *Parama Śiva* or *Saṃvit* and is said to be absent (*śūnya*) of objective being. In the *Śrī Yantra*, this primordial being is known as the *Mahābindu* as is represented by a dot at the middle of the yantra. So both the *bindu* and the original cosmic principle can be shown to have relationship to *śūnya*, either as a number or a concept.

DISCUSSION

The representation or symbol of zero, as discussed, seems to have arisen in a number of cultures where emptiness or absence is seen as an entity or force in itself. Whether this had anything to do with the development of the representation of zero is difficult to say. Whatever the case, *śūnya* provided the integral missing part to cosmologies with ideas of non-duality.

Representing the single Ultimate Reality with the numeral one is somewhat inappropriate as it is polar to –1. The number one implies actually the existence of two, (0, 1). The number, in the cosmic sense, is more appropriate to the first cause or the first arising as it has positive energy. On the other hand, zero has neutral energy; 0 – 0 = 0 and represents the union and thus, absence of opposites. In cosmological systems which see all things changing in cycles, *śūnya* represents both the beginning and end. We all start from the void and return there in the end (1 – 1 = 0). The representation of *śūnya* as a circle is then highly appropriate. Like time that moves in cycles such as the sidereal day, the lunar phases, the solar year etc., the cycles of nature also move in circular motion starting and ending at the same point.

Whether the *śūnya* symbol was originally intended to represent the empty space left in place value computation, it happened to be appropriate to other cosmic meanings naturally attributable to the number zero. The circle represents infinity, both by its form and by the space at its centre. What could be more appropriate for representing infinity than the cycle of time which repeated itself without fail and which always returned to the same beginning. Even the motion of the heavenly bodies, which produce the seasons, suggests such a form. The number zero also represents the cycle as any number subtracted by itself returns to zero. Thus, the 15 days of the bright phases of the Moon, in which it becomes progressively brighter, subracted by the 15 days of the dark phases, in which it becomes progressively darker, brings us back to the beginning. By representing the cycle, *śūnya* also symbolizes infinity. *Śūnya* is the only number that can be added or subtracted without changing value. If we think of infinity to include all numbers, then this makes sense. If one adds or subtracts a number to the entire set of numbers, all positive and negative numbers cancel themselves out. The result is the same when using *śūnya*, thus *śūnya* = (whole set of numbers), since :

if $a - b = c$
and $a - d = c$,
Then $b = d$

In addition to symbolizing infinity, *śūnya* can also represent the finite. The concept of limits is implicit in *śūnya*. Let us take a graph that shows two straight lines intersecting. The point of intersection represents *śūnya* for both the horizontal and vertical lines. The right side of the horizontal line and left side of the vertical line represent positive numbers, while the opposite sides of each line represent negative numbers. If we consider the positive side of the line approaching the intersection, the distance will become smaller as it approaches this point. Mathematically, one could continue to divide the distance infinitely as it neared the point and it would seem the line would never reach the intersection. However, there is a limit that must occur allowing the line to reach zero and then cross over to the negative side of the line. So, here *śūnya* represents the limit from which the distance can no longer be divided into a smaller unit. In Euclidean geometry, the intersecting straight lines keep going in a straight line into infinity. However, in the finite universe of non-Euclidean geometry, the lines curve back around meeting again on the same plane as the first intersection point. In this case *śūnya* represents limits at both the beginning and the end, another example of the excellent symbolism of the circle for *śūnya*. By representing both the infinite and finite, *śūnya* becomes all inclusive. In this sense, it is not surprising that esoteric thinkers even equated it with the all-encompassing, all-pervading deity.

Śūnya's symbolism as the point of origin, the "stuff" out of which things arise and even as the generator of things becomes apparent also when represented graphically. The point of beginning is always zero, no matter which direction one travels and whether it is in a straight, curved or crooked path. In some cases this point is mutual, while in others it is independent, but the point of origin can always be classified as *śūnya*. In relative terms, the point of ending can also be viewed as *śūnya* whether or not the beginning and ending are at identical points. Even when there is no ending, we can also view this as *śūnya* if we take all things as having polar opposites, since the whole set of infinite things cancels out to zero.

The importance of *śūnya* is, of course, greater in cultures where emptiness takes on an important role in spiritual thought. It provides the much needed symbolism not offered by the rest of the set of numbers. Obviously, *śūnya* is the symbol for those holding nihilist and similar views. Here it can represent the nihilistic extinction of things. However, historically *śūnya* offered its greatest reward to those systems possessing a philosophy of divine emptiness and voidness.

REFERENCES

Dasgupta, Shashibhusan, *An Introduction to Tāntric Buddhism*, foreword by Herbert V. Guenther, Berkeley, Calif.: Shambhala, 1974.

Diop, Cheikh Anta, *Civilization or Barbarism*, Chicago, 1991.

Escultura, E., *Diophantus, Introduction to Mathematical Philosophy*, Manila, 1993.

Gillings, Richard, J, *Mathematics in the Time of the Pharaohs*, New York: London, 1982.

Glass, Newman Robert, *Working Emptiness : Toward a Third Reading of Emptiness in Buddhism and Postmodern Thought,* Atlanta, 1995.

Khanna, Madhu, *Yantra*, London, 1979.

Kramer, Samuel Noah, *Sumerian Mythology : A Study of Spiritual and Literary Achievement in the Third Millennium* B.C., Westport, 1988.

Mookerjee, Ajit, *Tantra Art, its Philosophy and Physics*, New Delhi, 1994.

Nishitani, Kenji, *Religion and Nothingness*, trans. by Jan Van Bragt. Berkeley, 1982.

Rao, S. Balachandra, *Indian Mathematics and Astronomy: Some Landsmarks*, Bangalore, 1994.

Śūnyavāda, the Mādhyamika Thought, Dept. of Buddhist Studies, University of Delhi, Delhi, 1993.

Tola, Fernando, *On Voidness : Study on Buddhist Nihilism*, Delhi, 1995.

Wilhelm, Richard, *The I-Chung, or Book of Changes*, trans. by Cary F. Baynes, Princeton, 1990.

Word and Image in Maya Culture : Explorations in Language, Writing and Representation, Salt Lake City, 1989.

Zimmer, H., *Myths and Symbols in Indian Art and Civilization,* Washington, 1946.

NOW—Dimensionless (No-Mind-Śūnya) State

S.C. MALIK

Normally, the NOW is considered to be the 'Present' or just what is not any longer when it is. NOW is usually considered as part of time, by physicists as well as by philosophers; for the former it is no more than the subjective characterization of a particular instant of time; and for the latter the function of the word "now" is reduced to the instant which is simultaneous to the utterance of the corresponding word. This is another way of saying that it is part of time.

Clearly, the NOW is self-elusive: as soon as it is defined, characterized, localised, or merely taken as an object of awareness, it is likely to be no more really NOW. This is the paradoxical feature of NOW, an entity which both plays a role in temporal relations and does not belong to any time series. The "actual now" supports the argument against the "reality" of time and NOW is involved in many self-referential situations. Such an involvement is the key point which enables one to go beyond the traditional viewpoints in physics about time; and the temporal predicates used in ordinary language : IS - present, past or future; and, has been, is, or will be.

The plea of this paper, against a brief background of physics and philosophy, is that experientially the mind in its wholeness is the NOW, No-thingness. It is a spaceless, timeless dimensionless point that has only contact with seriality at right angles to it. It is not a fleeting moment; it is the noumenon through which everything is potentially present, projected and intuited. It is the sole line of communication between our enveloping totality and our apparent existence as separate individual creatures.

NOW alone informs us that we are in an apparent world in which we are not phenomenon only. We are nothing but the NOW, neither existing nor not existing. The NOW is a dimensionless point that throbs moment to moment, each moment being a totality. It is not apprehended by a system that reasons dualistically and is limited to three-dimension areas. It is apprehended WITHIN, whereby the concepts of time, space, linearity become subservient to it. Beyond thought, NOW is the SILENCE, a NO-MIND (ŚŪNYA) state, beyond knowing and not knowing; absence and presence. IT IS.

The Void is not of the nature of a black abyss or a bottomless pit.
Rather, its nature is vast and expansive like space itself. It is apprehended as serene, marvellous and pure, brilliant and all inclusive.
Above all does it partake of the nature of light and is not anything.
For Void is Mind itself; and Mind Itself is Void; that is the answer.
A Chinese saying.

PHILOSOPHY

NOW is usually considered as part of time, by physicists as well as by philosophers. Most physicists would say that it is no more than the subjective characterization of a particular instant of time. As for philosophers of analytic tradition, they would rather insist on the function of the word "now" in ordinary language, by ascribing it a so-called token-reflexive definition. According to them, the sentence "The event is happening now" means "the event is simultaneous to the utterance of the word 'now' which belongs to this sentence". Thus, NOW is reduced to the instant which is simultaneous to the utterance of the corresponding word, which is another way of saying that it is part of time.

However, this projection of NOW in time leads on to lose a non-negligible fraction of the content of the word. For, after all, one of the most striking components of its meaning is self-elusiveness: as soon as NOW is defined, characterized, localised, or merely taken as an object of awareness, it is likely to be no more really now. This is the well known paradoxical feature of NOW; an entity which both plays a role in temporal relations and does not belong to any time series: the "actual now" supports against the "reality" of time. NOW is involved in many self-referential situations. "Such an involvement is the key point enabling one to go beyond the traditional opposition between a 'static' and a 'kinematic' view of time". In ordinary language there is a subtle distinction that is maintained between temporal predicates: (is) present, past (past or future) on the one hand and temporal copulas (has been, is, or will be) on the other hand.

The relevant point to note is that the token-reflexive definition of 'now', however, is able to encompass the practical a posteriori content of the word, and fails to grasp the singularity of the concept of an 'actual now'. A now which cannot be referred to without becoming part of the field of description of 'another' now which then holds its role. The reason why this actual 'now' is so rarely considered is probably right because it cannot even be made objective without evacuating from it most of its original meaning. This boundary of thought has always raised a lot of interest from philosophy, ranging from Aristotle to Hegel, Sartre, Heidegger and other modern thinkers. Aristotle retained the experienced immutable and self-elusive aspects of the 'actual now'. Others think that these difficulties cannot be overcome and avoid mentioning any notion of the 'now' because it is something outside the time series, in relational terms. In short, the 'actual now' is necessarily, by its own definition absent from any tensed sentence. But the way one deals with this absence is of uttermost importance for understanding the role the absent entity plays in our account of time. One could keep silent about it but this would mean its denial, or use it in terms of extensive quotation marks denoting external higher level and making it explicit both in its absence and the actual role of the 'now'. Does it not show that the "actual now" might become a unifying concept in the analysis of the most paradoxical features of time? In short, philosophically, the NOW is just what is not any longer when it is.

PHYSICS

By contrast, in physics there is no expression of this extreme instability, since it has inherited from Paramenides a very strong tendency towards the formulation of invariants independent from any personal, spatial and temporal point of view. It was fundamental for physics to concentrate on an ambition to convey the primacy of immutability. This effort "culminated with Minowski's geometry of space-time; and with Einstein's Parmenidean characterization of the flow of time as an illusion". It is not surprising that the concept of NOW is missing in physics.

Actually, things are much more intricate even though the elimination of any mention of 'Now' may be a good strategy for the construction of a theory, because it is only at this cost that certain general requirements of symmetry can be fulfilled. But some kind of connection between the formalism and present actuality must be maintained in order to allow for empirical tests. In classical theories; and in special relativity as well, this connection is indirect, for it is performed through the intermediary of objective events.

In pure unitary quantum mechanics, however, objective events have no straightforward symbolic equivalent, but rather to the probability that such and such events happen. After all, actuality does not require any preliminary process of actualisation; nor does it imply that any "objectification" has taken place at a definite instant of the past. One could perfectly conceive that "Actualisation (if any) occurs now, always now". Quantum mechanics has forced us to come back to the point where the said traditional division clearly appears as a half-consciously instituted and historically moving boundary. The dualistic conception which puts a static, purely ordinal, model of time on the objective side and changes taking place Now on the subjective side, looks quite outmoded. Hence, attempts have been made to go beyond the dualistic view of time (past event and future event), by considering something like a "transcendental time".

MIND AND BODY

In view of this it has been suggested that a reflection on Now may throw new light on the mind-body problem. These investigations suggest a reversal of the usual order of questions about Now and time. The efficiency of the whole thought process would be improved if the phenomenological distinction between transcendental time and intentionally aimed at (physical) time, was replaced by a pragmatic distinction between the presupposed presence and the denoted instants. For instance, the distinction between 'I' and 'me' is a model; the former being the transcendental self, while the latter being the objective social self. The parallel may be between the 'I' and the missing Now; and between 'me' and the referred to instant of vocal utterance of the sound Now. All this leads to the corresponding statement: "The real Now is a presupposition of speech, even though it cannot be spoken about. The taken-reflexive now can be spoken about, but it does not deserve to be called Now". In short, the 'I' always therefore appears embedded somehow within the real Now, whereas the 'me' has the status of a partial feature of what is going on at the instant of vocal utterance of the word 'now'. Thus in quantum mechanics, the problem of solipsism, of consciousness, of presence and of actuality have been found so deeply interwoven.

In the here and now, the here is localised. But the actual now and its content cannot be specified now. This may be like the ego which is not easily located but at least definitionally in its normal accepted sense, it includes one's past history, my body etc. The actual now has no such aspects, even though it is the referring 'now' from which everything is referred to. There seems to be a relationship between the 'actual now' and the concept of a knowing subject. "...the idea of time flow is but the mataphorical expression of this infnite regress which in turn originates from the unique characteristics of the concept of the 'actual now' with respect to self-reference, or self-referential regress".

The similarities between this notion of the mind and the 'actual now' become striking at this stage: both are the missing end of a self-referential regress, at any level of it. The actual now is the missing end of a series of points of reference that are specifications of the 'moment' at which a given tensed relation holds. Again, the mind is the missing end of a series of machines that give meta-description of the machine just below them. An even more fundamental similarity is their being both, by construction, unspecifiable. They are foreigners in the domain they rule. The mind does not belong to the world of things and the "actual now" does not belong to time. Their very absence generates similar dualist temptations. These analogies are not superficial. In fact, a limiting concept of mind has been directly derived from that of the 'actual now'. The identification of some of the deepest aspects of the mind to 'now' is by no means new in the history of human thought.

The notion of mind is generaly too restrictive, since it often means the knowledge of a personal history, identity, feelings, as well as univocal association with a material substrate - human body - and its behaviour, usually giving rise to the idea of a private experience. But these aspects belong to the aspects of the known rather than to that of the knowing. The mixture of features which genuinely belong to the knowing subject with those which rather belong to known objects, in the common idea of the mind, is likely to be the main source of the difficulties discussed under the heading 'mind-body problem'. Conversely, replacing the border between mind and body by a border between knowing (mind in the restricted sense) and known (world) may yield a renewed appraisal of the whole problem. It may also afford a clear understanding of what mystical thinkers mean when they speak of the impersonality and timelessness of mind. The boundary between the known and knowing, is not sharp and the contents of each is very wide and arbitrary, i.e., the inner and outer boundaries are flexible. This is to suggest motion in the motionless present; while time flow has no purely objective foundation, it is not subjective in the sense of being restricted to some particular feelings of particular human beings, or even on the basis of some features of the human species. It only arises from what is both the most irreducible constraint of knowledge and the consequence of the mere possibility for it to exist. It arises from a universal condition for something to be known; neither objective, nor subjective, it rather arises from a characteristic (i.e., incompleteness) which is logically prior to the splitting of the known into 'moving' and 'motionless' or into an 'internal' and an 'external' field.

TIME

There are two ways to approach the enigma of time: to consider time as property or ever creation of the human mind, or to behold time as a frame of order effective for all events and objects. "...a Cartesian point of view that everything that exists is either mental or physical...the dualistic view is to separate matter from mind within a static conceptual frame; the fallacy of the monist is to look for a space-time description of mind..".. The trinity of mind, matter, time should be treated together rather than severed. Then the description of 'observer' and 'observed' provides the customary basis of scientific method; the classical method is fundamental for its success. But quantum theory corrects this point of view, because it says that the definition of an object may be an approximation only even if the frame of reference, the context or universe of discourse is clear. But this is true in principle only. Connected to all this is the problem of definition of the observer itself. But the classical theories do not take this into account since it implies that the observer has to know the true values of all propositions, including about his own state, i.e., it is a question of self-referentiality. This however admits incompleteness in terms of logical consistency. This leads to an infinite regress, reflecting that the act of observer and observed cannot be part of a universal theory, but is a meta-theoretical choice. While this will always remain a problem, speaking for physics, what one can do is to reconsider way of conceptualisation.

Beyond Duality

One way out is to go beyond the dominant western paradigm of opposites, if the dynamic nature of time is to be dealt with. But we always think of time in time; and even go beyond in time when considering change, temporal succession which is both mental and physical time - the former being inner and the latter being outer. However, this distinction between the inner and outer is not so distinct; the relationship may be a mutually dependent one, a twofold way - discrete time and continuous time. This may be a significant feature of brain activity and self-awareness; it is the self-referential nature of consciousness. However, the greatest link between the two times turns out to be status of the NOW. It may be the NOW which is the window to a dynamics beyond temporal succession. Physiologically also it suggests that there are zones of simultaniety or directional temporal zones. Similarly, physics moves beyond the monolith idea of classical picture, reflecting the dynamic nature of the continuity of time, i.e. the actuality of time is embedded in a NOW - speaking at a meta-level, being the cardinal point in continuous and discrete time structures, between external and internal time, may lie the meaning of "fact" that there is a basic complementarity of space-time description and causality.

There is a whole range of history of theories in physics related to time, notions of time and in philosophy about NOW, about mental functions. The variable of time is fundamental in physics but the problem of 'passage' of time is the most problematic, since there is no way to measure this passage; the processual or dynamic aspect of time is frozen in classical physics. The instant "now", time has not been stated, nor described, such as the fact there is a symbol for zero, of the infinity and of the negation. Attempts have been made to formalise this moment, in terms of its productions, the process of actualization and in terms of causality, everything originates from the present and not the usual explanation of causality from past origins etc. The flow of time does not exist by itself, the methodology is through operational mediations and not through some physical apparatus but more through language. Nevertheless, it is not a position of test, as it refers to the temporal context of its own production. NOW is at the same time a bench-mark; and an instant which is unstable in a sense, but nevertheless it is in perpetual displacement i.e., with respect to itself.

Thus an instant (*a*) is a transition between itself and another, (*b*) it is an intuitive statement which has been attempted to be formalized in mathematical statements of Combinatory Logic or perhaps hypersets. For example, a written text is just a structure statistically laid down on its medium; it takes a meaning only in so far as a reader brings it up to the universe of meaning. In this way "meaning" is the same as "creation of meaning", a statement by which we see its self-referential nature. Time: is the production of NOW. The philosophy underlying the self-referential model of time rests on a twofold time: the constructed "physical" time and the transcendental time of the transition *a* to *b,* i.e., time flow at different speeds depending on whether the signifying register is language, music, 'annoyance', 'subjective' and 'objective' time.

Thus, existence is not made up of one instant, but a series of instants—motion being within the instant, not a derived quantity, self-referentially related to the start and the end of an instant. But this is nonstratified logic, where one of the important characteristic of the relation $a = a$ to b is that the term a is at the same an object, entering the transition a to b, that the operation itself, since a to $b = a$: it is the transition, an interval and also the starting point of the transition.

So, NOW while it escapes physics, it necessarily involves the dimension of meaning, together with its necessary underlying transcendence. The self-referential structure of temporality can be extended by the use of a generalized self-reference, leading to a non-linear time which can help to understand meta-psychological notions such as the timelessness of the unconscious. In short, NOW can be understood by a simultaneous exploration of physics, philosophy and meta-psychology, because it lies at the crossroads of time. The intention is not to reduce the description of brain processes to physics or mathematics. Instead, a reversal is suggested. A shift of focus from physics to biology, from objects to functions should be useful to develop a conceptual frame embracing matter and mind.

Conceptualisation is inevitable in any act of knowing. But the knower and the known should not be considered as two independent metaphysical entities. They come to be together. Perception of time and creation of time constitute an inseparable unity. This self-referentiality should be appreciated as the ultimate expressions of the subordination of time to objective categories of being. One can, thanks to the temporal aspects of drive and pleasure, which are the border between mind and body, claim that inversely, their link is temporality itself. In the interaction between time and Gestalt, a whole world of plurality, a plenty of extended presences are enclosed. From their interactions continuous time is abstracted. Seen from "outside", from external time, extended presence should appear as discrete time. With a present sensation, "inner" extended presence is reached. The past sensation or fact is the Gestalt enclosing a present sensation. The "dividing line" between presence and past is observation. Note that observation is defined by no change. In a world of changes it seems to be at a standstill; it seems to be nowhere. In a world of sensations and moved by sensations, observation can be ascribed a point-like existence only.

In contrast, mind is pure dynamics. It does not exist. Existence is an attribute of a space-time appearance. The objects of mind are concepts. Objects are static. This is the basis of the temptation to forget dynamics. However, mind as dynamics can re-enter a static world and dissolve it.

Earlier, two complementary concepts of time were defined, i.e., discrete and continuous time. The bridge between both is the central notion of a conceptualization based on categories of change: re-entrance. Only re-entrance creates time in the sense of temporal sequence. Extended presence becomes fact — or memory-like in a chain of succession. Creation of time is the measurement of a difference by itself via re-entrance. Res cogitans has to be described as extension. But not as extension in res extensa (space-time). Being in extended presence, there are no sensations. There is no connection to "external" time. To abstract continuous time, the basis of entrance has to be "forgotten". Being is space-time, the observer cannot perceive his own Gestalt. William James said, "Consciousness is an unconscious act".

"Within standard quantum mechanics, (there are) no 'really existing' facts". The emphasis is on standard and existing facts; and it is not univocal and inescapable conclusion about facts as such since all theories must have facts. What it means is that fact is a definitive statement, or it is a performative one, or it is expressing some thing reflexive, about what the speaker is seeing or knowing.

The latter may be called phenomenal facts, since it deals with what is presently experienced by the speaker, i.e. first person experience; and/

or spatio-temporal indexial statements, viz. "I am experiencing thunder now". But this statement is abstracted from the present experience, even though the character of immediacy is perfectly in evidence; yet this cannot be communicated except as an act of faith, through the belief that my memory is reliable; the distinction is between having seen from the certainty of seeing it.

Sciences like physics deal with this type of phenomenal facts. But in this internal relational structure of phenomenal facts, the main difference from corresponding statements of classical theories is that the boundary between the condition and the conditioned has shifted. The condition has to encompass of what is called the outcome and not only the preparation.

Moving from the phenomenal fact to objective fact, beyond solipsist account, is forming law-like descriptions, e.g. even if the lightening is not seen or heard, given such and such conditions the possibility and predictability is there though not necessarily so since every possible condition and variable cannot be taken into account. But this is the realist's position, in terms of third person entities, as public—objective—entities. This again in terms of intersubjective communication cannot be carried out except as a possibility and an agreement of sorts, in general terms. In fact what I am experiencing now, I may not be able to communicate to myself a little later on. There may be traces of evidence of this experience, for which phenomenological fact and even legally speaking, I have to find ways to say of the certainty of having seen something, but also to show a permanent evidence—exhibit to substantiate the claim. Objective fact would thus be a logical, constructed and accepted version of the experiencing which happened at that instant, its properties. But since I am the only eye-witness, it has to be accepted as a matter of faith. This would be true even if there were co-witnesses, since more often than not different versions are stated for the same event - whatever definitions one gives for this word. The event has existence because of its properties located relationally spatio-temporally. Nevertheless, this is still in terms of possibilities and probability statements, nothing permanent about these statements and these are as if statements in the plural world of quantum mechanics, in contrast to classical physics. Property means something inherent of that object and the idea that given similar conditions it is reproducible. Properties may be alterable but can continue to be traced to some historical origins.

In short, within the paradigm of pure unitary quantum mechanics, there is no general concept of property, no spatio-temporally continuant objects left, no exhibit to show, and consequently no ground on which to base any discourse about objective facts.

However, in spite of all this there is room for structures. In its very structure quantum mechanics allows for inter-subjective agreement, i.e., there are conditioned statements about phenomenal facts, whereby at least the status of the phenomenal fact remains. This is a pragmatic fact, in the sense that it allows in its quantum account of the event, though it does not solve all difficulties, nor speaks of permanent properties of objects as such. This assumes that quantum mechanics is a universal paradigm. It is not easy to speak of potentiality to actualization of an event or a fact. The transition cannot be assessed except arbitrarily; or in the statement NOW which is really atemporal since time location cannot be ascribed to it, since the moment one utters the word NOW —even a chronological time — it is gone; thus, facts are really about past events, or future events instants. The Now, being indivisible, is already in the past while being spoken of and apprehended. Analogously, the I, first person, when spoken of is also not apprehended immediately; it is spoken of as a third person, much the same way one speaks of the NOW.

There is thus reason to renounce the concept of objective fact, the concept of event, and the third-person definition of NOW(as a spurious present). One can thus state that actualisation does not occur in time, but it is not true that it does not therefore never occur. In short, thus, actualisation occurs NOW and always NOW; and NOW is not a time, but nor is it never. However, actualisation is not to be taken to be synonymous with "objectification", i e., transition from potential facts to an actual objective fact; this atemporal transaction does not have a "when" at its end. The question 'when is NOW?' has no answer.

The NOW has to be a recorded NOW, for it to occur and be communicated. However, like the subjective self, which when describing becomes a third person objective fact, the atemporal knowing subject is certain, it is not knowable, but the known and knower are not in the now. Maybe a distinction has to be made between the transcendental ego and the empirical or psychological one, of the atemporal now and the pragmatic now. This is like being in the world and being not 'of it', as opposed to 'for it'. Maybe the knower and known are one, the property of the knower being the known, or known is the knower and there are no two entities; like, the rays of the sun and the sun are the same since without the rays what is the sun? The transcendental ego coexists with those of the others and is not individual and particular in nature.

In short, there is a transcendental field within which the world is posited, in which the existence of objects is relative only. It would be like knowledge of and knowledge for, distinction; suggesting theoretical and practical contexts. However, it suggests that there is incompleteness of knowledge, which has a reflexive, semantic and pragmatic side to it. Bearing this in mind, the paradox diffuses, if these distinctions are kept in mind, bearing in mind also the principle of complementarity.

In both Eastern and Western philosophy, it is the problem of the relation of an "external reality" to preceptual experience and hence "awareness". The only difference from classical physics, to quantum mechanics and relativity, is that once for all it breaks with the tradition whereby the "external reality" is considered as a pale show (colourless, odourless, noiseless etc.) of the world of commonsense realism.

SUMMARY AND CONCLUSION

In a statement like "I am eating", means consciousness as "I" (as Self) comes into existence "am" (a psychosomatic form), becomes a process of "eating" or identifies with a name, label etc., i.e. I is eating in an existential-experiential moment of Now, moment to moment, but not as an abstract conceptual existence. It is in this sense that the so-called paths, each being Universal Consciousness are whole, else total. Each like the grain of sand or drop of water, is complete and whole in its essence and is part of what we may call a perfect system and is perfect in its esential nature, out of time and space concepts, eternally or in the eternal moments of Now.

In this sense, all are part of nature or cosmos (meaning order), beings are already in harmony, in tune. This natural continuum is shared by all inanimate and animate, past, present and future in the Now, in what is. It is a state of communion with a uniqueness all its own — there is a state of freedom existing by itself — not from anything. It is beyond space and time, beyond cause-effect and other notions. What is required is change of paradigm, a change of direction, about our very perceptions. For example, consciousness has to be apprehended in terms of each and is exactly where we are, Here and Now.

The Now is dimensionless-point that vibrates and throbs in moments of Now, NOW, each moment being a totality. The current systems are limited to three-dimensionless areas, whereas the total view belongs to another dimension, not necessarily a fourth one or may be one which is not related and does not at all arise out of the other three. The fourth dimension includes the other three, once it is apprehended, the three are subservient to the fourth and serve the 'whole' mind and reflect it—where the unsaid and unseen comes forth, beyond words.

Again, such understanding or apperception of wholeness, is not objective — my understanding is valueless to you. It cannot be transferred or transmitted to you. It can only be pointed out. It does imply the cessation of any kind of mentation, implying a vivid state of consciousness free of all abiding discursive, mental activity by means of images, forms and figures. This is non-objective, impersonal awareness — beyond words, or thought of the personal self. The reference is to the subjective total mind in the non-dual sense which is often referred to as one-mind, no-mind etc. All that is needed to de-condition one's self, by waking up. The need is to strip oneself of all garments for only in total nakedness will that come which we seek.

What is this Now, the total Now, this dimensionless point? The Now has no duration, it is not horizontal. It has only contact with seriality at right angles to it. Its measurement is within, in this timeless dimension so to speak. It is not another extension of time, it is eternally present in the Now. This is what links us to the other three dimensions. Yet, it is not a fleeting moment; for in it exist all things potentially and as manifestations which is time, i.e. from the noumenon everything is projected. It is the invisible portal through which intuition reaches us. It is that universal or limitless interior which is all we ever were or will be, and which is 'out of time'. It is the sole line of communication between our enveloping totality and our apparent existence as separate individual creatures. Now alone represents that we are in an apparent world in which we are not, since therein we are appearances — phenomenon only.

We are nothing but as the present, neither existing nor not existing. That dimensionless point is the eternal silence, more eloquent than words yet from which words appear and into which they disappear. Silence is not a negation of noise, it is a stillness. It is a duration beyond duration since in itself it is atemporal and infinite. Knowing this, its incalculable potency becomes apparent, for this is no other than whole mind. This, there is nothing to be acquired, only the error to be exposed. The mind in its wholeness is Silence, beyond opposites and comparisons. This may be referred to as the Universal Presence; the impersonality of pure consciousness, inaccessible to the process of identification, represents the plenitude of the void, space, the one-mind. And the Universal Presence is at the same time Universal Absence, for there is nothing to be present and nowhere for a presence to be. And whosoever is conscious of this and nothing else, is that eternal Now. Knowing, understanding this, abolishes one. The dream vanishes and only non-manifestation-timeless, spaceless, the immutable—remains. He lives in that eternal silence.

REFERENCES

Artigiani, Robert, 1990, "Thermodynamics and History", in Kishore Gandhi, (ed.) *The Odyssey of Science, Culture and Consciousness*, New Delhi.

Bateson, Gregory, 1984, *Mind and Nature - A Necessary Unity*, London.

Bitbol, Michel and Eva Ruhnau, 1994, *Now, Time and Quantum Mechanics*, Editions Frontiers, France.

Bohm, David, 1980, *Wholeness and the Implicate Order*, London.

Bohm, David and Krishnamurti, J., 1985, *The Ending of Time*, New York.

Cassidy, David C., 1990, *Uncertainty : The Life and Science of Werner Heisenberg*.

DeChardin,Teilhad, P., 1959, *The Phenomenon of Man*, New York.

Dirac, P.A.M., 1958, *The Principle of Quantum Mechanics*, O.U.P.

Dossey, Larry, 1982, *Space, Time and Medicine, London.*

Ferris, Timokthy, 1991, *The Mind's Sky; Human Intelligence in a Cosmic Context*, Bantam.

Fiumara, Gemma Corradi, 1990, *The Other Side of Language - A Philosophy of Listening*, Routledge, London and New Yok.

Gandhi, K. (ed.), 1990, *The Odyssey of Science, Culture and Consciousness*, New Delhi.

Harris, Errol E., 1991, *Cosmos and Anthropos - A philosophical interpretation of the anthropic cosmological principle*, New Jersey, London.

Heisenberg, Werner, 1959, *Physics and Philosophy* - Glifford Lectures, George Allen and Unwin, London.

Kothari, D.S., 1986, "Science, Culture and Values", Mimeographed.

Kothari, D.S., 1989, "Complementarity Principles : Physics and Beyond", in *N.R.L. Technical Bulletin*, special issue, New Delhi.

Kothari, D.S., 1990, "The Perception of Truth in Science and Philosophy", in Gandhi, op. cit.

Lester-Smith, E., 1975, *Intelligence Came First*, Illinois.

Lovelock, J., 1979 *Gaia : A New Look at Life on Earth*, Oxford.

Lovelock, J., 1988, *The Ages of Gaia*, London.
Lovelock, J., 1989, *Modern Civilization—A Crisis of Fragmentation*, New Delhi.
Malik, S.C., 1994 (ed. with Saraswati, B.N. and Khanna, Madhu), *Art: The Integral Vision*, D.K. Printworld, New Delhi.
Malik, S.C. (ed.), 1995, *Intercultural Dialogue and the Human Image, Maurice Friedman at the IGNCA*, New Delhi.
Malik, S.C., 1995, *Reconceptualising the Sciences and the Humanities—An Integral Approach*, Manohar and NMML, New Delhi.
Prigogine, I., 1980, *From Being to Becoming*, San Francisco.
Prigogine, I., 1984, *Order out of Chaos*, New York.
Schroedinger, E., 1948, *What is Life*, Camb. U.P.
Sheldrake, R., 1980, *A New Science of Life*, London.
Weber, Renee, 1986, *Dialogue with Sages and Scientists*, New York.

Śūnya: A Holistic View

R. SATHYANARAYANA

Śūnya (skt.) is derivable from *śūna* which is formed from the root *'śvi'* (cl. 1, *'bhvādi'*, *parasmaipadi*) also taking the forms *'śū'* or even *'śvā'*. The root means 'to swell' (morbidly), 'to enlarge'; and by semantic extension, 'hollow'. It is in this last sense that the word *'śūnya'* has now established itself to mean empty, void, vacant, vacuity, absence, free from, nonentity, nonexistence, cypher (zero), nought and space. These meanings may be conveniently collocated under the dual categories of space and matter-energy. Absence, nonperception or occupation of space is a condition for the appearance/perception of matter-energy; nonappearance/nonexistence of matter-energy is a condition for the appearnace/perception of space. Thus (epistemological question of absolute existence or of appearance-perception apart), *bhāva* and *abhāva* of space and of matter-energy seem to occupy conjugate positions in referential consciousness and thus offer a clue to its nature. Positivists and phenomenologists view human consciousness as structured to perceive only positive existence and hold that nonexistence of negation is only a linguistic problem.

An attempt is made in this paper to present *śūnya* in some of its aspects viz. number-nonnumber, void-full, nonexistence/nonappearance (*abhāva*), state of consciousness, subjective and referential space etc. from different disciplinal viewpoints in a holistic perspective.

NUMBER AND NON-NUMBER

Śūnya as zero is a Hindu contribution to world of mathematics. It is the basis of place-value and decimal system which have exclusively pervaded the number theory. It was disseminated into western mathematics through the Arabs; zerc and the numerals were propagated in Europe by the Italian Leonardo Fibonacci of Pisa.

Zero plays a dual role of both number and nonnumber in mathematics because it has properties of both. This duality is recognised in both the East and the West since early times. *Śūnya* has for its synonyms *kha, ākāśa gagana, ambara, chidra, bindu* and *randhra*. It may be noted that the last two suggest its mathematical sign of a dot or small circle, known in India and South East Asia for abcut 1,500 years, while the others are terms denoting space, thus meaning objectlessness.

Every Hindu work on mathematics devotes a chapter on zero, called *śūnyagaṇita*. Zero is neither positive nor negative; it is the border between the two, the sum of two equal but opposite (+ve and –ve) entities of the same kind. Its special properties are: when added to or subtracted from any number, the latter remains unchanged, i.e. $a \pm 0 = a$. The product of any number with zero is zero i.e. $a \times o = 0$. When any number is divided by zero, the quotient is mathematically indeterminate and is called *kha-hara* or *tac-cheda* ($a \div 0 = khahara$). The square and square root etc. of zero remain zero. Many of these properties are mentioned by Śrīpati in his *Siddhāntaśekhara* (4.6) (11th cent. A.D.)

vikāram āyānti dhanarkāni na śūnyasaṃyogaviyogatas tu/
śūnyād viśuddhaṃ svaṃ ṛṇaṃ kṣayaṃ svaṃ vadhādinā khaṃ khaharaṃ vibhaktam//

The functions of zero are clearly elaborated by Brahmagupta in his *Brāhmasphuṭasiddhānta* (18.30-35) in the 7th cent. wherein he mentions *taccheda* too.

Some ancient references by Indians (Hindus, Jainas and Buddhists) to zero may now be mentioned. *Ṛgveda* mentions *śūnam* (2.27.17; 2.28.11; 2.29.7; 3.33.13; 8.45.36) and *śūne* (1.105.3; 7.1.11; 10.37.6); *śūna* (from *'śvi'*) is mentioned in *Gopatha Brāhmaṇa* (1.3.9) while *'śūnyam'* is referred to in *Śatapatha Brāhmaṇa* (*Mādhyandina* , 2.3.1.9, *Kāśyapīya*: 1.3.1.4). *Taittirīya Brāhmaṇa* has *'śūnyāya'* (2.1.2.12). *Yajurveda* assumes place value and decimal systems when it mentions (17.2) numbers in a geometric progression in decuple order from 10^0 to 10^{12}. While explaining the tabulatory device called *prastāra*, Piṅgala mentions in his *Chandaḥśāstram* (c. 400-200 B.C.) (8.28, 31) *śūnya* explicitly. The Jaina canonical work

Anuyogadvāra (c. 100 B.C.) mentioned 29 decimal places. *Puliśasiddhānta* (= *Pauliśasiddhānta*), a work on astrology (c. A.D. 400) mentions that the sidereal revolutions in the four *yugas* run to ten decimal places. Āryabhaṭa I mentions zero in his *Ārybhaṭīya* (1.2) (A.D. 499) and Bhāskara elaborates the concept and its functions in his commentary on the work (2.2). The *Pañcasiddhāntikā* of Varāhamihira (c. A.D. 550) mentions *śūnya* as a number (1.17). The *Bṛhatkṣetrasamāsa* of Jinabhadragaṇi-kṣamāśramaṇa (A.D. 609) mentioins 13 decimal places (1.69, 71) while Siddhasenagaṇi (6th cent. A.D.) demonstrates a property of *śūnya* which is inhered in the decimal system while commenting on Umāsvāti's *Tattvārthādhigamasūtra* (3.11). Explaining the different, multiple roles of the same person as mother, daughter sister etc. Vyāsa gives the analogy of a *rekhā* (stroke) functioning variously in various decimal places, in his commentary on Patañjali's *Yogasūtras* (3.13). Vijñānabhikṣu in his *Yogavārtika* on this *bhāṣya* passage of Vyāsa emphasizes this role of *śūnya*. Śṃkara draws a similar analogy in his *Brahmasūtrabhāṣya* (2.2.17).

Sarvajña Someśvara III explicitly mentions in his *Rājamānasollāsa* (2.97) in A.D. 1131 the numbers 1, 2, 3, . . . 9 and the place value of zero.

ŚŪNYA AND INFINITY

Śūnya and *khahara* (= *ananta, taccheda*) may be regarded as mutually obverse entities in Hindu mathematics even as *śūnya* and *pūrṇatā* are regarded as obverse of each other in Hindu philosophical systems.

Brahmagupta stated that not only zero divided by any number is zero but that zero divided by zero is also zero. This latter view is no longer accepted today but the quotient is now regarded as indeterminate. Brahmagupta holds that the quotient of any number divided by zero is indeterminate, called *taccheda*, a descriptive term which means that (*tat*) which has zero as denominator (*cheda*). This is *khahara* which is synonymised by Bhāskara (12th cent. A.D.) with infinity in his *Bījagaṇita* (śl. 20) :

asmin vikāraḥ khahare rāśāv api praveṣṭeṣv api niḥsṛteṣu/
bahuṣv api syāl layasṛṣṭikāle 'nante' cyute bhūtagaṇeṣu yadvat//

Thus the *khahara* is a *vikāra* indicating an infinite (endless) quantity which remains unchanged when anything, however large, is added to it or subtracted from it, even as the Infinite and Immutable (Brahman) remains unchanged when countless material things (*bhūtagaṇa*) are added to it at the time of creation and are subtracted from it at the time of dissolution. There is thus a conceptual equilibrium between *khahara* and *brahman* through indeterminacy. This is echoed in the Upaniṣadic *śāntimantra* :

pūrṇam adaḥ pūram idaṃ pūrṇāt pūrṇam udacyate/
pūrṇasya pūrṇam ādāya pūrṇam evāvaśiṣyate//

ŚŪNYA AND INDETERMINACY

The number system of Western mathematics including zero is based on the following five postulates enunciated by Peano in A.D. 1899.

i. Zero is a number, ii. Successor of any number is a number, iii. No two numbers have the same successor, iv. Zero is not a successor of any number, and v. Every property of zero, also belonging to the successor of any number that has the property belongs to all the numbers.

Paradoxes of infinity such as the one mentioned by Bhāskara are discovered by Dedekind, Richards and Bertrand Russel. Certain mathematical operations involving zero give ambiguous and imponderable results and are called indeterminates. They are interpreted by methods of Differential Calculus. The Creationist school of mathematics holds that numbers and their mathematical relationships are creations of human thought and therefore provide an insight into the working of the human mind, especially of referential consciousness. So an encounter with irresolvable mathematical contradictions and paradoxes would lead to a profound discovery about the nature of the human mind. Instances of undecidable mathematical theorems include Fermat's Last Theorem :

$$a^n + b^n = c^n \text{ (for } n > 2)$$

and Goldbach's conjecture that every prime number (> 2) is equal to the sum of two integers.

Gödel's discovery regarding undecidability of mathematical formal statements is very relevant to the concept of indeterminacy: Can *both* a formal mathematical statement S and its negation ~S be proved? If this were true of *any* pair S and ~S, it would prove catastrophic because *any* mathematical statement could be proved. Is there any statement S such that *both* S and ~ S cannot be proved (i.e. S is undecidable)? In 1930, Kurt Gödel used Hilbert's methods of formal axiomatics and demonstrated that for *any* mathematical system as complex as the theory of infinite sets or of numbers,

1. it is impossible to prove the consistency of the axioms and
2. it is possible to generate undecidable axioms from the axioms.

Finally, when mathematics extends the use of numbers to realms beyond experience of the real world (e.g. operations involving zero) or follows their implications too far and ends up in paradoxes and contradictions, it is worth-while asking : is it the world that lacks consistency, or is it the mind that lacks consistency?

ZERO AS PHYSICAL ENTITY

Following are some of the physical connotations of *śūnya* as zero:

1. It is the point of line marked on any graduated scale as beginning of reckoning, e.g. in a thermometer, barometer, ammeter, voltmeter etc.
2. It is the absolute beginning of anything.
3. Absolute zero degree is that state of matter in which the thermal motion of molecules is nil.
4. Zero-point energy of a substance is the kinetic energy remaining in it at zero degree absolute. According to quantum mechanics, a simple harmonic oscillator does not have a stationary state of zero kinetic energy. This is in accordance with the Principle of Uncertainty; for, a particle cannot be at absolute rest at the point of origin. Such residual energy is distributed to the normal modes of lattice vibration, involved as the binding energy of the crystal.
5. Zero entropy is defined as the entropy of a system in equilibrium at zero degree absolute, according to the Third Law of Thermodynamics.
6. If zero is synonymised with *abhāva* (negation) or nonapprearance of matter-energy, it is then reduced to mere space-time. If warping (curvature) of space-time is a condition for the appearance of matter-energy, is it possible to apprehend the former without reference to an object?
7. In modern physics, the distinction between matter-energy and the sorrounding space is no longer tenable. Force-field Theory, developed by the application of geometry of space (*śūnya*) to gravitational field of subatomic particles, holds that the boundary between a material particle and the space surrounding it is no longer sharp. Void (*śūnya*) is now

regarded as a dynamic quantity of the highest importance. Matter may be thought of as constituted by regions of space in which the field is extremely intense, i.e. field is the condition for a charged particle. This relation between field and charged particle readily recalls to the mind the metaphysical relationship between *kṣetrajña* and *kṣetra* as taught in the *Bhagavadgīta* (13.1.34). The analogy of the Force-field and charged particle will be invoked later in the context of *śūnya* and art experience (Section 8).

ŚŪNYA AND THE ABSOLUTE (KEVALA)

Just as void and matter are in the final analysis indistinguishable, *śūnya* is imponderable and is beyond the dialectical opposites of being and non-being. This is the Absolute. This notion is as old as the *Ṛgveda* (10.129.1).

nāsad āsīn no sad āsīt tadānīṃ nāsīd rajo no vyomā paro yat/
kim āvarīvaḥ kuha kasya śarmann ambhaḥ kim āsīd gahanaṃ gabhīraṃ//

(Then there was neither being nor nonbeing, no earth, no intervening space, nor the heavenly worlds. What then surrounded it, what was its substrate?)

This thought is echoed in some Principal and Minor *Upaniṣads*:

asad vā saumya idam agra āsīt (*Tai. Up.* 2.7.1)
yo 'yoniṃ yonimadhitiṣṭhtyeko .... śāntimatyantameti (*Śve. Up.* 4.11)

(He neither is nor is not. He—*Śiva*—is nothing but peaceful.)

Other references are from *Maitri-Upaniṣat: śuddahaḥ pūtaḥ śāntaḥ* (4.6.28; 7.4.) *śūnyabhūto niḥśabdaḥ* (6.2.3); (He is pure, holy, void and peaceful; He is of the nature of void and silence). *Nṛsimha-uttaratāpinī Upaniṣad*; ānandaghanaṃ śūnyam abhavat ātmaprakāśaṃ śūnyam (6) (He is intensified bliss and void; he is void and self-effulgent), *Amṛtabindūpaniṣad: śūnyaṃ kṛtvā nivātmakaṃ śūnyabhāvaḥ* (10, 11) *Tejobindūpaniṣad : sarvaṃ ca paramaṃ śūnyaṃ, śūnyabhāvaḥ (2, 11).*

All Creation proceeds from *Śūnya or nirguṇa brahman,* is sustained by it and returns to it: *Yatovā imāni bhūtāni jāyante yena jātāni jīvanti yat prayānty abhisaṃviśanti (Tait. Up. 3-1)*

Thus the referential universe emerges from the nonreferential absolute, — which is beyond being and nonbeing and is empty — and finally re-enters it.

This Hindu vision of the void has a Chinese parallel: According to Chung Tsai, the Great void cannot but consist of *ch'i*; this *ch'i* cannot but become dispersed so as to form once more the great void. *Ch'i* is conceived in ancient China somewhat similar to the modern quantum field, as a tenuous and non-perceptible form of matter which is present throughout space and may condense into solid material objects. According to Taoism, Tao is infinite and endless creativity. It is totally empty (*śūnya*). The Tao of heaven is empty and formless according to Kuan Chung (7th cent. B.C.). According to Lao Tzu, Tao is a hollow valley, a vessel which is for ever empty and thus has a potential for containing an infinity of things. (In this way both the Hindus and the Chinese regard void and fullness, as mutually obverse and complemental). A metaphor may be drawn for this from subatomic particles.

Such particles are not static and permanent, but are dynamic and transient, coming into being and vanishing in a ceaseless dance of movement and energy. This recalls to the mind an imagery from Śaṃkara :

punar api jananaṃ punar api maraṇaṃ punar api jananījaṭhare ....
(Bhajagovindastotram)

The *jīva* in its sojourn of cycles of birth and death is like the subatomic particle which flashes forth from the womb of the Primordial Absolute and dances its way back into the same abode. This eternal dance in the matrix of the being and nonbeing transcends into the Absolute in the image of Naṭarāja, King of Dancers.

ŚŪNYA IN METAPHYSICAL THOUGHT

Śūnya may be synonymised with *abhāva,* i.e. nonbeing, nonexistence or negation. This is an important category in all Indian-Hindu or otherwise-metaphysical systems.

Abhāva is broadly classified into (a) *saṃsargābhāva* and (b) *anyo'nyābhāva*. The former is again of three kinds: *prāg-abhāva*, *pradhvaṃsābhāva* and *atyantābhāva. Prāgabhāva* is prior nonexistence, e.g. nonbeing of curds in milk; *pradhvaṃsābhāva* is nonexistence after destruction or death, e.g. a jar after destruction, a person after his death; *atyantābhāva* is absolute nonexistence such as the sky flower, hare's horns or son of a sterile woman. *Anyo'nyābhāva* is mutual negation as the nonexistence of cow in horse and *vice versa.*

Indian philosophical systems are preoccupied largely with proof of non-existence, i.e. *abhāva pramāṇa* while their western counterparts are mostly concerned with nature of nonexistence (*abhāva svabhāva*). As already mentioned, phenomenologists and positivists among the latter reject the theory of negation and regard it as a linguistic problem rather than philosophical. They consider that the human mind is structured to receive and understand only positive existence.

Abhāva in Pūrvamīmāṃsā

Kumārila Bhaṭṭa and his school aver that all discrimination between cause and effect and between different things is anchored on *abhāva.* Otherwise only confusion and absurdity result. Again, *abhāva* is real. If it were not, it would not have varieties, which are accepted on all hands.

According to this school, *abhāva* is a separate category; its proof is *anupalabdhi,* i.e. nonapprehension. Absence of knowledge is the means of knowledge of absence. This school postulates that a thing has both a positive and a negative form and sometimes it presents itself to consciousness in one form and at other times in the other, resulting in *bhāva* and *abhāva* respectively. *Abhāva* is related to space and time. Cognition of *abhāva* is not the result of direct perception (*pratyakṣa*), inference (*anumāna*) or presumptive testimony (*arthāpatti*).

A refinement was introduced into Kumārila's logic of *abhāva* by Śrīdhara (in his *Nyāyakandalī*) by saying that nonapprehension is only a *liṅga* (mark, probans) for cognition of nonexistence while *yogyānupalabdhi* (uncontradicted nonapprehension) is the source of such cognition.

However, according to Prabhākara and his school, *abhāva* is not a separate category and nonapprehension is not proof of cognition of *abhāva.* He holds that negative judgement is valid not because it has a corresponding negative fact but because it refers to a positive fact indirectly. So cognition of 'no jar' is not a positive knowledge of a negative fact but is the negative knowledge of a positive fact, viz. that we do not perceive the jar. This knowledge is preceded by an inferential process in which nonapprehension is probans.

Abhāva in the Nyāya School

The views of this school on *abhāva* are similar to those of the Bhāṭṭa school with an exception, viz. *abhāva* is cognized primarily as a unitary, unanalysable phenomenon by direct sensory perception (whereas, for example the inference of fire on the hill is mediated by the memory of *vyāpti* or concomitance). A question is raised: since *abhāva* has no colour or shape and has no direct sensory contact (*indriyasannikarṣa*), how can it be apprehended by direct sensory perception? This is answered with an analogy: atoms have colour and shape and are yet not seen by the eyes. Ether has direct contact with the eye and is yet not seen by it. Therefore the presence or absence of colour or shape is not sufficient condition for sensory perception.

Abhāva in the Vaiśeṣika School

The views of this school are notable largely because of the refutation of Buddhist views on *śūnya* or *abhāva*. Such refutation is made strongly by Vyomaśiva in his *Vyomavatī* which is commentary on Rāvaṇa's commentary on Praśastapāda's *bhāṣya* on *Vaiśeṣika Sūtras* of *Kaṇāda*:

Objection : (a) A particular *abhāva* coexists with a particular object at a given time and place and not with all objects at all places and times.
(b) If *abhāva* is producible (*utpatti*), how is it that it is produced by a positive entity? For a cause produces an effect of a *sajātīya* kind (of the same class or type).

Answer : It is observed that a single cause may produce various, different effects, e.g. a single source may produce different colours, tastes and others qualities.

Objection : If *vināśa* can be produced (*prāgabhāva*) or destroyed (*pradhvaṃsābhāva*) then the latter means the negative of a negative, i.e. a positive entity. So *abhāva* should result in *bhāva*.

Answer : Nature of *abhāva* is unlike that of *bhāva*. Further, *prāgabhāva* is not produced but is destroyed; *pradhvaṃsābhāva* is produced, not destroyed.

Śaṃkara on Abhāva

Śaṃkara is quite influential in refuting arguments on *abhāva*. Thus he shows that the *asatkāryavāda* of the Vaiśeṣika school, viz. that effect does not exist before it is born is not acceptable (*Brahmasūtrabhāṣya* 2.1.6.18). He refutes the view that *abhāva* can be a *hetu* (generating cause) for *bhāva* (existence), i.e. that nothing may produce something (*ibid.*, 2.2.4.26, 27). He also demonstrates that *abhāva* (nonexistence) cannot be a general cause from which specific things come into existence and if it were, there would be only chaos but no order in the real world (*ibid.*, 2.2.4.27). Nonexistence is absolute, as the son of a sterile woman and has no varieties. The notion of void as nonexistence or emptiness is totally unacceptable to him and he posits the notion of fullness and infinity as *brahman* in contrast. Rarely for him, he is vehement and somewhat harsh in denouncing the Buddha as wholly irrelevant and malevolent in misleading the gullible (*ibid.*, 2.2.5.32). He successfully attacks the *śūnyavāda*, especially of the *vijñānavāda* and *mādhyamika* schools (*ibid.*, 2.2.4.28-32) and does so again in his *bhāṣya* of *Māṇḍūkya Upaniṣad* (*Alātaśānti prakaraṇam*). He regards *ākāśa* (space) as an entity, not void.

Buddhism on Śūnya

Fundamental to Buddhism are not four *bhāvanās* postulated by the Buddha, viz. *kṣaṇika* (momentariness of consciousness, evanescence of perception), *svalakṣaṇa* (fundamental difference between particle and particle), *duḥkha* (sorrow) and *śūnya*. *Śūnya* is that which cannot be perceived as *sat* (existing), *asat* (nonexisting), *sadasat* (existing-nonexisting), *sadasam anubhayātmaka* (neither existing nor nonexisting). It is defined in the *Majjhima Nikāya* as not a nihilistic entity but a state of objectless consciousness (*animitta cetosamādhi*). Originally, *śūnya* referred in the context of *samādhi*, to the state of obstruction which accompanied increasing concentration of consciousness in meditative practices, a method favoured in Tantra, Mantra and Yoga practices too. This however developed later into soullessness (*anātmatattva*) in Abhidharma literature. The Mādhyamika school had developed the theory of *pratītyasamutpāda* according to which the individual soul evolves from the subtle to the gross through successive generation of twelve states, viz. *avidyā–saṃskāra- vijñāna- nāma- rūpa-ṣaḍāyatana- sparśa-vedanā- tṛṣṇā- upādāna-bhava jāti- jarā- maraṇa*. This was regarded as a state between the dialectical opposites of being and nonbeing. Nāgārjuna reinterpreted *pratītyasamutpāda* as *śūnya*.

The state of *śūnya* is realised through a nondual knowledge designated as *prajñā*; it is the illusory nature of all phenomena. *Prajñāpāramitā* is the state of illuminative nondual consciousness in which mind becomes non-mind and cannot be characterised as existing or nonexisting. Thus, the Absolute is imponderable and is beyond the dialectical poles of being and nonbeing. This recalls to the mind the Transcendental Absolute suggested in the *Nāsadīya* hymn of the *Ṛgveda* and in the *Śvetāśvatara Upaniṣat*.

Contrary to popular impression, *śūnyatā* in Buddhism means neither nihilism nor positivism but is absolutism. It manifests as a psychodynamic dimension. 'It does not aim to describe the world of appearance but attempts to revalue common experience. It is a critic of philosophical theories on common experience. It does not change the nature of things but realigns our vision of things.'

The *Svacchanda Tantra,* an ancient text of Kashmir Śaivism teaches (4.289-90) of seven *śūnyas:ūrdhvaśūnya, madhyaśūnya, adhośūnya, vyāpinīśūnya, samanāśūnya, unmanīśūnya* and *paramaśūnya*. The first six are impure and contain progressively weaker seeds of phenomenalisation and should be discarded to reach the seventh which is the Pure Being, Pure Consciousness, Supreme Reality and is totally *nirvikalpa*. The *Aṣṭasāhasrikāpiṇḍārtha* enunciates the doctrine of sixteen or eighteen *śūnyas* in the attainment of *nirvāṇa*.

In course of time, the notion of *śūnya* was transformed into that of *vajra* by the Vajrayāna school of Buddhism. Here (e.g. *Advayavajrasaṃgraha*, p. 37), *śūnyatā* which is firm, substantial, indivisble, impenetrable, impossible to burn and imperishable, is called *vajra*. The parallel between this description and that of Upaniṣadic Brahman is notable.

In fact, the *Caryāgītis* composed by Buddhist *siddhācāryas* between the 8th-12th cent. A.D. in a symbolic language (occasionally Sanskrit) which could have mothered Bengali, Oriya or Bihari, and described as a song-genre in musical treatises from the 12th to the 17th cent. A.D. reveal this parallelism. Their tantric-philosophical content swings through the *śūnyavāda* of the *mādhyamikas*, the positivism of Aśvaghoṣa and of the Vijñānavāda school of Maitreya, Asaṅga and Vasubandhu. They also reveal the ubiquitous influence of the monistic thought and of *brahmavāda* of the Upaniṣats.

Tantric Buddhism also speaks of the union (incomprehensible, inexpressible, inseparable) between *Śūnyatā* and *Karuṇā* (the female and male partners of the union respectively). Another metaphor of the same type speaks of *śakti* (spoken of as *śakti, avadhūtī* etc.) in the *caryāgītis* which has union (*yuganaddha*) after phased processes of purification with *sahajakāya* (= *vajrakāya*) and becomes Vajra herself. The symbolism of this metaphor of both the *Caryāgītis* and of the cultic practices has been considerably weakened frequently by a crass misinterpretation of the female partner as a corporeal woman.

The Vijñānavādins hold that *śūnyatā* is negation of knower and knowable. Nāgārjuna describes *śūnyatā* as experiences in four progressively abstract phases viz. *śūnya, atiśūnya, mahāśūnya* and *sarvaśūnya*, which differ from each other in both cause and effect. Among these, *śūnya* is synonymised with *strī, vāmā, chandramaṇḍala-padma* and the first vowel '*a*'. It is *āloka*, i.e. Light and is defined as *prajñā* in which the *citta* (referential or sensorial consciousness) is dependent (on other things than itself, *paratantra*) and contains 33 impurities (*doṣas*). *Atiśūnya* is called *upāya* (means), *dakṣiṇa sūryamaṇḍala, vajra,* or *avidyā*. It shines like rays of the moon. This phase is designated as *parikalpita* (constructive imagination). It has 40 mental functions. *Mahāśūnya* is also called *ālokopalabdhi*. It is the third stage of *citta* in its ascent to *sarvaśūnyatā*. It is of the absolute (unrelated) nature (*pariniṣpanna*). It is contaminated with seven impure mental functions of defilement such as forgetfulness. From these three stages follow 160 principles of impurities which have *prāṇa* for their vehicle; *sarvaśūnya* is wholly pure, self-illuminant. It is the state of purified knowledge, absolute truth, supreme omniscience and is beyond all duality. It is called *śūnyasamādhi*.

Jainism and Śūnya

As far as I am aware, Jainism does not admit *abhāva* or *śūnya* as a category into its philosophical system. But an ancient legend relating to the first Tīrthaṅkara, Ṛṣabhadeva sheds interesting light on the Jaina concept of zero as a number and nonnumber. This is narrated by Kumudendu in his *Śrī Bhūvalaya*, a celebrated work in the *saṅgatya* metre composed in the early 9th century A.D. According to its introductory section, Ṛṣabha the emperor had two wives, Nandā and Sunandā from whom he begot two sons, Bharata and Bāhubali and two daughters, Brāhmī and Sundarī. He bequeathed his whole empire to Bharata (who thus became the first Indian emperor after whom the country is named) and the small township of Paudanapura to Bāhubali. Then his two young daughters besought their share of ancestral wealth. (V) Ṛṣabha had already given away all his material wealth to his sons. He then decided to bequeath the incomparably precious spiritual and *śāstra* wisdom to them. So he seated Brāhmī on his left lap, Sundarī on his right lap, divided zero (*śūnya/pūrṇa*) into two halves, wrote one half on the left palm of Brāhmī with his right thumb and the other half on the right palm of Sundarī with his left thumb. He showed Brāhmī that her share evolved graphically into 64 phonemes of the Kannaḍa language (which comprehend fully those of Sanskrit). This formed the base from which the totality of all verbal/literal expression in all the *śāstra* and non-*śāstra* disciplines in the (then) known 18 major and 700 minor languages (*samasta śābdāgamaśāstra*). Sixty-four is a mystic/magical number in Jaina mythology. He then showed Sundarī that her share of the zero evolved (graphically) into the numerals 1, 2 . . . 9. Thus both the qualitative and quantitative aspects of knowledge were comprehended. He then proceeded to encode both into a single system in which the number (*aṅka*) and the syllable (*akṣara*) are so related that both may be reduced to numerical tabular forms of 64 rows and 64 columns from which any desired branch of knowledge may be decoded by resorting to particular *prastāra* procedures. This is called *akṣarāṅka, samanvaya paddhati* and was taught by Bāhubali to Bharata in the battlefield in the course of a mere *muhūrta*. *Śrī Bhūvalaya* is an amazing work, hailed as the eighth wonder of the world, in which different *prastāra* procedures are adopted to decode the number tables into incredible and diverse literary results. Like the two hands which gave and received the two complementary halves of zero, both branches of knowledge are mutually complementary and are obverse of each and are taught as *mokṣa śāstras*. Thus what emerges from zero/full and constitutes the entire referential cosmos, exists and finally returns to zero/full. Both the points of emergence and return are thus indeterminate, infinite, void/full. There is thus an interesting parallel to the Upaniṣadic thought

pūrṇam adaḥ .... pūrṇam evāvaśiṣyate / yato vā imāni .... abhisaṃviśanti/

Yoga and Śūnya

Patañjali mentions the term *śūnya* on four different occasions in his *Yogasūtras*:

(i) *śabdajñānānupāti vastuśūnyo vikalpaḥ* (1.9)

Here *vikalpa* is one of the five modalities of the mind, which is not predicated referentially (i.e. on an object) but falls back upon a linguistic net. Thus, it operates through associative ideation created by words around an object even though the object *per se* does not exist (*vastuśūnya*), e.g. when we say that the *puruṣa* (Supreme Being) is of the nature of (consciousness) energy, the two are really congruent; so which qualifies which?

(ii) smṛtipariśuddhau svarūpaśūnyevārtha–mātra–nirbhāsa–nirvitarkā (1.43)

This aphorism speaks of a mental state called *nirvitarka* wherein deliberation is relinquished to attain a balanced state of super deliberation. It is attained by purifying memory by freeing it of deliberation of word symbols and inferences. Then consciousness assumes merely the shadow of the intended object and becomes devoid of itself (*śvarūpaśūnya*). This balanced state then brightens into conscious knowledge as the intended object and nothing more. This is the super deliberative state (*nirvitarka*).

(iii) tadevārtha–mātra–nirbhāsaṃ svarūpaśūnyam iva samādhiḥ (3.3)

Patañjali teaches here that *dhyāna* (contemplation) itself becomes *samādhi* when the former (intense focus on the presented idea, i.e., *pratyaya)* shines forth in consciousness as only the intended object and nothing else and is as it were, totally (otherwise) void (*svarupaśūnya*). When the three, viz. fixed attention (*dhāraṇā*), contemplation (*dhyāna*) and concentration (*samādhi*) become instruments with a single objective, it is called *saṃyama* which is the means of attaining of various *vibhūtis* (super normal manifestations).

(iv) puruṣārtha–śūnyānāṃ guṇānāṃ pratiprasavaḥ kaivalyaṃ svarūpa–pratiṣṭhā vā citiśaktir iti (4.34)

This is the final *yogasūtra*, teaches the culmination of *yoga sādhanā* and describes *kaivalya*. This is pure consciousness-energy abiding wholly in itself. This is reached when the three *guṇas* (*sattva, rajas* and *tamas*) whose essence is causes and effects are inversely generated (reversed into their source) and when all human goals of achievement are voided (*puruṣārtha-śūnya*)

This nonreferential, objectless, self-abiding consciousness, variously described as *mokṣa, kaivalya, svarūpajñāna* etc. is the ultimate objective of all forms of spiritual quest of the Hindus such as *yoga, mantra, tantra* etc.

Patañjali speaks of *abhāva* on four occasions. Of these, two are in relation to the sleep phenomenon (1.10) and for removal of obstacles to spiritual quest by self-study (*svādhyāya* consisting of *praṇava-japa* and its contemplation). The other two relate to *kaivalya:* when *avidyā* ceases to exist, the *puruṣa* is delinked from intelligence (*buddhi*), hence bondage ceases. This is *kaivalya* in which the *puruṣa* is no longer contaminated with the *guṇas* which are the generating cause for *duḥkha* (2.25). An important milestone in yoga practice for *kaivalya* is the obviation of subconscious impressions (*vāsanā*). These must be removed from generating cause (*hetu*), consequence (*phala*), mental substrate (*āśraya*) and stimulus (*ālambana*). Among these, *hetu* is a ceaselesly revolving six-spoked wheel of rebirths, motivated by *avidyā*. Its six spokes are: the three dualities: *dharma* (righteous living), *adharma* (nonrighteous living), *sukha* (pleasure, deriving from *dharma*), *duḥkha* (pain, resulting from *adharma*) *rāga* (attraction, resulting from *sukha*) and *dveṣa* (repulsion, deriving from *duḥkha*). These again generate righteous living etc. *Phala* accrues from the *hetu*. When unfulfilled, mind becomes the substrate for subsconscious impression. When a thing confronted with some object phenomenolises any subconscious impression in itself, then that object is the stimulus of that *vāsanā* (4.3).

Daharākāsa and Śūnya

Most spiritual disciplines in India share a common cosmogonic concept, viz. Transcendental Reality manifests at two extremes, cognized as the microcosm and the macrocosm and that the microcosm is the miniature of the macrocosm. These two cosmos are like the most extreme points on the circumference of the circle (which are also the most proximate points there). And the circle is the manifestation and proliferation of the selfsame, single point, viz. the centre (or *bindu*). These extremes of apparent reality are enveloped by zero (circle, void/ full) and infinity (coalescence of two circles), both of which are indeterminate (*kha* and *khahara*).

The forgoing cosmogonic concept is extended by the Hindus in a metaphysical context to space, which is the condition for all Appearance. According to the Upaniṣads, there are two forms of space: external, elemental, vast, macrocosmic, called *bhūtākāśa*; internal, subjective, minute, microcosmic, called *daharākāśa*. This latter is enveloped in the heart lotus *hṛdaya puṇḍarīka* in the city (= Body) of Brahman, the *brahmapurī*. *Daharākāśa* is *śūnya*, the abode of *brahman*; nay, the very *brahman*.

Daharākāśa is mentioned in such diverse scriptural sources as the *Kātyāyana Śrautasūtta* (14), the *Śatapatha Brāhmaṇa* (10.6.3.2) and the *Kena Upaniṣat* (2.1). The *Chāndogya Upaniṣad* devotes a whole chapter to it (1.8) and develops an entire spiritual *śāstra* called *daharavidyā* in self-realisation. *Bādarāyaṇa* devotes an entire section (*adhikaraṇa*) to *daharavidyā* (1.3.5.14-21) in his *Brahmasūtras*. Śaṃkara holds an elaborate debate on whether the space mentioned here refers to the elemental space (*bhūtākāśa*), discriminatory consciousness (*vijñānātmā*) or to the individual soul (*jīva*) and, refuting these views, holds forth that the *daharākāśa* is *paramātman* (*ibid*., loc. cit.).

It is thus clear that this minute, inner space, *daharākāśa* is the abode of pure consciousness, *nirguṇa brahman*, i.e. the condition of disappearance, while the vast, objective, macrocosmic, outer space, *bhūtākāśa* is the condition for appearance, i.e. the state of referential consciousness or *saguṇa brahman*. Each *ākāśa* is *śūnya*, not in the sense of *physical void*, but in the sense of *spiritual fullness*.

Tantra and Śūnya

In Tantra *śāstra* Ultimate Reality is held to be pure consciousness which is called *cit* or *saṃvit*. This is nondual, unlimited by mind, body, name or form and is of the nature of *sat-cit-ānanda*, i.e. existence-consciousness-bliss. This is symbolised as I-consciousness of Śiva whose power is Śakti, nondifferent from Śiva. The Śiva aspect of Consciousness is immutable whereas the Śakti aspect manifests as mind-body, name-form, the plurality of the referential world, the apparent reality. These two aspects are respectively called *Prakāśa* and *Vimarśa*.

Such Pure Consciousness is often called *Śūnya* in this *śāstra*—not in the sense of negation, but of fullness, of infinity, of intellectual or verbal indeterminacy. Various procedures are taught from yoga, mantra, āgama and other disciplines to realise this as Silence—*niḥśabdam brahma ucyate*—through Sound, in which the practitioner passes through various phases of *śūnya* to attain the highest and ultimate, called *mahāśūnya*. These are known as *śūnya, atiśūnya* and *śūnyātiśūnya*.

The practitioner of *Śāktatantra*, i.e., *Śrīvidyā* encounters *śūnya* in both theory and practice in different levels and modes. Only a few instances may be mentioned here. The *Śrīcakra* is a symbolic device representing the forcefield of consciousness - energy which may be manipulated and guided along disciplined pathways in each of several psychodynamic levels in decreasing degrees of grossness (or referentiality) or increasing levels of subtlety (or abstraction) in which the ultimate goal is absolute, pure, nonreferential or objectless consciousness. This goal is the point at the centre called *bindu*, the *mahāśūnya* which is attained by a process of centration; concentration leads to involution of consciousness whereas excentration leads to evolution. This is somewhat analogous to the process in which the outermost, differentiating electron of an atom, with all four quantum numbers different and revolving on its own axis at a finite distance from the nucleus, and is thus a differentiated individual, gradually loses the appropriate quanta of energy at various levels (comparable to degrees of referentiality) and is finally drawn into the nucleus (the ultimate, undifferentiated consciousness). This is called *saṃhārakrama*. The reverse is true in *ṣṛṣṭi karma,* i.e. evolution into apparent reality.

The practitioner of *Śrīvidyā mantra* encounters *śūnya* in an audio–visual psychodynamic process. After his consciousness transcends to a phase of abstraction above the three levels of decreasing referentiality (called *kūṭas*) and he begins meditation on the peak of the third *kūṭa* (viz. *hrīm* or the *Śākta praṇava*) consciousness is progressively concentrated (i.e. its volume is diminished) in nine phases (commencing from '*m*' in *hrīm*) in each of which it assumes at the same time a form of a nonreferential sound as well as a form of nonreferential light. The sound is experienced as more and more subtle and the light, more and more intense as the consciousness becomes more and more concentrated. That is, it moves inwards and culminates in silence and in total self-effulgence or *prakāśa*. At each phase, the subtlety of sound and intensity of light are directly proportional to the diminution of consciousness volume in a spatiotemporal reference frame. During this practice, the *sādhaka* experiences five gaps or *śūnyas* comparable to the central circle in a peacock feather in hue and shape, located after each of the first five sound-light forms, commencing from the centre of the forehead. The sixth and final, called *mahāśūnya* occurs in *unmanā* (the phase transcending mind) in total silence and total light above the *brahmarandhra* and *bindu* in the thousand-petalled lotus. This corresponds to the experience of the *bindu* in *Śrīcakra*.

Again, in *Kuṇḍalini Yoga* of *Śrīvidyā Tantra*, the *sādhaka* arouses the

Kuṇḍalinī energy to induce her to enter the *brahmanāḍī* (*suṣumnā*) which lies within the hollow (*śūnya*) of *citriṇī nāḍī*. This *nāḍī* is strung like gems with the six *yogacakras* lying in the *vajradaṇḍa* (vertebral column). *Kuṇḍalini* flows through this *śūnya*, pierces the pericarps of the *cakras*, reaches the *bindu* (*mahāśūnya*) of the *sahasrāra* (thousand-petalled) lotus, and there becomes one with Ādideva, pure, universal, absolute consciousness. These instances contain esoteric details which must be learnt and experienced only at the grace and guidance of the Guru.

Pratyabhijñā and Śūnya

Trika Śāstra is a recension of *Śaivāgama* which originated and flourished in Kashmir. It is so named because it is the philosophy of the triad: Śiva, Śakti, Nara, or *para - parāpara - apara* and exists in three categories, viz. Āgama, Spanda and Pratyabhijñā. It comprehends the concept of *śūnya* in two contexts: refutation of the Buddhist notion of *śūnya* and as an instrument for attaining to the *śivabhāva*, but is not the absolute, divine consciousness. The *ātman* is neither body (*Cārvākas*), nor *prāṇa* (Vaidikas), nor *buddhi* (Yogācāra Buddhists) nor *śūnya* (Mādhyamika Buddhists) according to the *Śivasūtras* (1.1). It is *vyoman* or *ākāśa* which is devoid of all difference. It is synonymous with *svāpa* (deep sleep) in which all objective phenomena are absent (*ibid*., 1.15). *Śūnya pradhāna* is explained as the inner soul of the subtle form (*puryaṣṭaka*) that has a contracted manifestation with the main constituent (*pradhāna*) being either void or *prāṇa* (*ibid*., 3.10).

The *Spandakārikā* teaches that *śūnya* cannot be conceived without Light serving as its support (comm. on śl. 2). It refutes Nāgārjuna's contention that *śūnya* is devoid of all supports (*ālambana*), of all constitutive principles (*tattvas*) and of residual traces of all *kleśas*. This view is acceptable only if it is contemplated on the substratum of consciousness- bliss. In the *trika* system the highest consciousness is free from all spatiotemporal and other notions.

Abhinavagupta in his *Vivaraṇa* commentary on the *Parātrīśikā* confirms that *śūnya* is *vyoman* or *ākāśa*. He explains *svadeha* in *śvadehāva-lakṣaṇam* as embracing *prāṇa, puryaṣṭaka, śūnya* etc. When *śūnya* is the fourth progressive state of compactness/solidification adopted by *Śivabīja*, it is objectless consciousness. The third *brahman* (= the syllable 1) is *vyoman* or *śūnya* which is the inner state of the four elements, viz. earth, water, fire and air. Devī asks Śiva: How does the unsurpassable divine consciousness (*anuttara*) bring about immediately the achievements of the identity of the empirical-I with the perfect I-consciousness of Śiva so that one acquires sameness with *Khecarī* (universal consciousness-power)? One anwer to this question is : *anuttara* may be decomposed into *anuttara*. The former means *ākāśa* = *śūnya, khecarī* = *khe caratīti*. Here *kha* is the nonreferential, universal I-consciousness. All referential existence or appearance is the concretisation of consciousness, much as a charged particle in a forcefield is the intensification of the field at that point, as matter-energy is the warping of space-time. The Buddhist *Hevajra Tantra* offers (13.2) a theory for the phased process of emergence of the concrete from the abstract, of the concrete sound from the *śūnya*:

prathamaṃ śūnyatābodhaṃ dvitīyaṃ bījasaṃgraham/
tṛtīyaṃ bimbaniṣpattiḥ caturthaṃ nyāsam akṣaram//

The Pratyabhijñā system classifies, as in *Parātrīśika* (and its *Vivaraṇa*) experients into *śūnyapramātā* and *dehapramātā*. Between these, the *śūnyapramātā* has only life, but not even breath; he experiences only the void, has no-feeling in the sensorium and has no sense of objectivity. The soul lies in sleep due to *moha* for a long time in both the *pralayakevalin* and *śūnyapramātā*. However, in the *śūnyapramātā* the vital breaths *prāṇa* and *apāna* continue to move subtly till they lie unconscious, in *pralaya*. The *sañcita* (accumulated) *karmas* of both, nevertheless, occur as *saṃskāras* in this state.

Nātha and Vīaśaiva on Śūnya

The nine *Nātha siddhas* Matsyendra, Gorakha etc. have taught the *śūnya* principle in the pursuit of self-realisation. Gorakhanātha queries his Guru Matsyendranātha: 'Sire, whence does Light emerge? Whence does air come forth, whence the essence of the three worlds? How does one cross the ocean of the phenomenal world?' The Guru replies: From *śūnya* proceeds everything: into *śūnya* everything returns. The mind finds peace in *śūnya*.' Gorakhānath avers in his *Gorakhabāṇi*: Know the *śūnya* to be *Śabdaśiva* (*paripūrṇa, brahman*); in it is *nirvāṇa*: this can be found in your own body; learn it from the Guru. *Śūnyasamādhi* is mother; father is *nirañjana brahman*; the mind becomes firm and unwavering from it. The ultimate state of consciousness transcends the dialectic opposites of dwelling and void, being and nonbeing, *sat* and *asat*. It is both, it is neither.

The *bindu* whispers incessantly in an inverted well of *śūnya* in the subtle *nāḍi*: 'I am That! I am That!'

The Ultimate is Light : *śūnya* is its shrine; sound, its door; light, its icon; the flame, boundless form; formless mind, *the unmanī* (beyond mind.)

The Nātha and Vīraśaiva religious practices and principles have many parallels, the latter probably being influenced by the earlier former. These similarities are very pronounced in their two foremost leaders, viz. Gorakha and Allama Prabhu both of whom hailed from Karnataka. Both systems worship attributeless, formless Śiva; their final goal is the attainment of perfect union between Śiva and Śakti (*samarasa*). Principles of *Sahaja* and *śūnya* are common to both. Neither recognizes icon worship and pilgrimage. Guru is inestimable, highest, in both. The practice and principle of Nātha engender the *ṣaṭcakra* while those of Vīraśaivism embrace *ṣaṭ-sthala*, both serving as staircase for spiritual ascent by the arousal and leading of *kuṇḍalinī*. Body-discipline and *bhakti* are respectively their *modi operandi*. Gorakhanātha and Allama are their foremost exponents and are regarded by their followers as Śiva Incarnate. Both employ spiritual conundrums as media for verbal expression and explicate the concepts of *sahaja, śūnya, nirañjana, kadali, sirigiri, nāda, bindu, ṣaṭcakra, ummanī mūlādhāra . . . sahasrāra, suṣumnā* in exactly the same way in the same contexts. Gorakanātha and Basava also employ a similar conceptual framework and imagery, *śūnya* figuring in both.

Śūnya is, according to Allama, the ultimate goal of human endeavour, the ultimate principle and abode of final rest. It is called *śūnya liṅga* or *niṣkala liṅga*. He bases this view on the authority of Śaiva Tantra :

> *ādimadhyāntaśūnyaṃ vyomā vyomavivarjitam/*
> *dhyānajñānadvayor hīnaṃ śūnyaliṅgam anāśrayam//*

Elsewhere, it is said

> *antaḥ śūnyaṃ bahiḥ śūnyaṃ śūnyā śūnya daśā diśaḥ/*
> *sarvaśūnyaṃ nirākāraṃ nirdvandvaṃ paramaṃ padam//*

Śūnya is the state of absolute, nonreferential consciousness, constant, nondual, incomparable, indivisible, *sat-cit-ānanda*. Its experience is *śūnya samādhi:*

> *ūrdhvaśūnyaṃ adhaḥ śūnyaṃ madhyaśūnyaṃ nirāmayam/*
> *sarvaśūnyam anābhyāsaṃ samādhes tasya lakṣaṇaṃ//*

Śūnya has no form, no attribute. It is Light, it is beyond knowledge, beyond dualities. It is silence, stillness, impenetrable. It is the seed or the egg from which creation emerges. It is into this that the *sādhaka* endeavours to render his body, mind, thought and feeling into states of *śūnya* and finally dissolves them progressively. It is beyond the *ṣaṭsthala*. In short, it is *Brahman* as taught in the Upaniṣads.

Vīraśaiva cosmogony is mystic and assigns an important role to *śūnya* in the evolution of the cosmos: such evolution proceeds broadly in three progressive phases, viz. *śūnyaliṅga, mahāliṅga* and *sādākhyā*. *Śūnyaliṅga* is again of three subphases: *sarvaśūnya - nirālamba, śūnya liṅga* and *niṣkala liṅga*. The first of these is a state which preceded creation. *Śūnyaliṅga* has been briefly described above. *Niṣkalaliṅga* is *satcidānanda brahman*, the source of all categories and principles of creation and is the primordial cause. The march of evolution in the successive phases from the subtle to the gross is studied in some detail by Allama and others. The notion of *śūnya* plays an important part in the final liberation in Vīraśaiva dogma and is different from the *nirvāṇa* of Buddhism. The latter is attained through negation of all appearance and of the internal causes which generate it. On the other hand the *śūnya* of Vīraśaivism (and the Upaniṣads) is positive; it comprehends everything and is of the nature of fullness and perfection. *Śūnya* and *pūrṇa* are thus merely obverse of each other and transcend the dialectical opposites of existence and nonexistence.

Śūnya and Pūrṇa

Pūrṇa is derivable from the etymons '*pṛ*' and '*pur*' meaning sustenance, fullness, self-sufficiency or self-containedness. In the Vedas and the Upaniṣads it connotes totality or wholeness of the absolute, fullness, completeness or conclusion.

The *Atharvaveda* (6.80.1) describes the fullness of the full moon which is akin to the pervasiveness of *śūnya* and of the wholeness. In yet another passage (10.8.79), it employs the word *pūrṇa* which leads to the concept of supreme being as the embodiment of fullness. The *Bṛhadāraṇayaka Upaniṣad* (5.1.1) has the famous passage that if fullness is taken away from fullness, only fullness remains and that nirguṇa brahman is full; saguṇa brahman is also full. The *Yogavāsiṣṭha* (3.10.28, 29) expatiates on this synonymity of the Absolute with Fullness. It makes the significant statement: The unborn (precreation) universe is full and the born universe is also full. The *Haṭhayoga–pradīpikā* (4.56) describes the pervasiveness of fullness. These and similar ideas in the sacred lore of the Hindus are very comparable to those of *śūnya* discussed in the foregoing.

Zero symbolises both *śūnya* and *pūrṇa*. Two coalescent zeros symbolise infinity; this is neither coincidental nor accidental. This suggests that *śūnya* and *pūrṇa* are complementary aspects of the self-same infinity.

ŚŪNYA AND COGNITION

Space and Time are fundamental categories in physics. They are equally fundamental to all metaphysical thought. It has been mentioned above that space as *śūnya* is the condition or substrate for Appearance; likewise, Time is the substrate for Change. Therefore Space and Time are substrate for all cognition. They may be objective or subjective. According to the Relativity Theory, objective space and time are inseparable, interdependent parts of a single whole. A consequence of Mach's Principle is, as pointed out by Einstein in the formulation of the General Theory of Relativity, the mutual conjugability of space (void, *śūnya*) and matter (full). This may be called *deśa śūnya*. *Kāla śūnya* may be regarded as the nonexistence of linear time. Therefore *deśa-kāla-śūnya* is the nonexistence of referential experience. This corresponds to absolute existence of only the Force-field, i.e. of charge, but without a charged particle which may be considered as a local intensification or concentration of force.

On the other hand, subjective space and time are cognised in many aspects, as Casirrer shows (*The Philosophy of Semiotic Forms*, vol. 3, p. 646) : subjective space, intuitive, mythical, symbolic, sensuous, conceptual, expressive, significative, empirical, real, absolute, mathematical, geometric, kinesthetic, natural and formal space. Similarly time is cognised variously as subjective time, experienced, intuitive, metaphysical, mathematical, biological, objective, absolute, historical and physical time.

ART, CREATIVITY AND ŚŪNYATĀ

The creative force charging consciousness is called *pratibhā*. For this reason it is called *śakti* (force). It is defined as consciousness which causes innovation or creativity to bloom or flash ever afresh. In metaphysical parlance, it is visualisation or intuition of supra sensory knowledge which transcends intellectual analysis. In this scene, it is intense mental focus, and relates to *samādhi*. In the aesthetic context, this is a close kin of *brahmānanda*. It is different from other states of consciousness because while recollection (*smṛti*) relates to the past, *mati* to the future, *buddhi* to the present, *prajña* relates to all three at once. It functions both at a general level as nonreferential void (*samādhi*) and at a specific level of consciousness in creating or innovating something which has not existed before.

The Force-field theory may be extended to explain creativity in art. Kṣetra (Force-field) is analogous to consciousness which is charged with *pratibhā*, *kṣetrajña* to the artist. Just as in the physical field the distinction between the charged particle and the field surrounding it dissolves, in the aesthetic field also the distinction between I-consciousness of the artist and of the *pratibhā*-loaded consciousness disappears. All referentiality falls away from consciousness of the artist leaving behind a condition of spatiotemporal vacuum (*deśakālaśūnya*) which is pervaded only by the charged field of *pratibhā*. This is the state of intense concentration or intensification without localisation of specific co-ordinates or parameters and is analogous to *samādhi*. In the next phase, *smṛti, mati* and *buddhi* make their appearance, localising or condensing the intensity of the field with the specificities of grammar, idiom and technique of the particular art-mode. In the next phase, just as the localisation of the field intensity leads to the definition or concretisation of the charged particle in a space-time continuum, in art the processes of articulation and expression set in. This aspect or function of *pratibhā* is called *kārayitrī*. Again, in the Force-field, the effects of a charged particle are felt only by another charged particle because both coexist within the same medium. Similarly in art, the analogues of the two charged particles are the artist and the connoisseur. Both exist in the Force-field of *pratibhā*. The effect, viz. *rasa* of one charged entity, viz. the artist is felt by the other charged entity, viz. the connoisseur because of the common medium of generalisation and abstraction—*sādhāraṇīkaraṇa* — of the psychodynamic experiential content of the situation. This aspect of *pratibhā* is called *bhāvayitrī*. In this stage the experiential parameters which are peculiar and specific to the artist and to the connoisseur fall away in a process called *sāmānādhikaraṇya* (equalisation, equiliberation), so that only the *summum bonum* of the art experience viz. *rasa* is experienced by both.

The *śūnya* in which the art experience takes place is not one of negation but is of fullness. It is a negation of only referentiality; it is a visualisation of supra sensory fullness and wholeness which baffles intellect and recollection.

CONCLUSION

An attempt is made in this paper to consider *śūnya* from multiple viewpoints as a mathematical, physical, mataphysical, mystical, philosophical, tantrik, subjective, objective and aesthetic entity. It emerges at two strata of human mind, viz. as a concept and as an experience. As a concept it remains concealed in the cave of the unknown, as an imponderable, indeterminate entity comparable to Vāk:

catvāri vāk parimitā padāni tāni vidur brahmaṇā ye maṇīṣiṇaḥ/
guhā trīṇi nihita neṅgayanti turīyaṃ vaco manuṣyā vadanti//

(ṚV 1.164.45)

It has four parts, all of which are known to the Knowers of Brahman; three are concealed (from Man) in the (dark) cave of the unknown. Man speaks only the fourth part.

And this leads to nought at one end and infinity at the other (both being indeterminate, unknown), embracing everything in between.

As an experience, it is void of only the referential or phenomenal. All creation proceeds from it, is sustained by it and returns to it. Yet it is beyond them. It neither is, nor is not. It transcends all duality. It is universal, absolute. It is pure consciousness.